OPTIMIZATION ALGORITHMS - EXAMPLES

Edited by **Jan Valdman**

Optimization Algorithms - Examples
http://dx.doi.org/10.5772/intechopen.71370
Edited by Jan Valdman

Contributors

Orhan Kurt, Zheng Hong Zhu, Gefei Shi, Honggui Han, Lu Zhang, Junfei Qiao, Constantin Udriste, Henri Bonnel, Ionel Tevy, Ali Sapeeh Rasheed, Vadim Shmyrev, Jian Gao, Hamidou Tembine, Yida Xu, Julian Barreiro-Gomez, Massa Ndong, Michail Smyrnakis, Andrzej Łodziński

Notice

First published in London, United Kingdom, 2018 by IntechOpen
IntechOpen is the global imprint of INTECHOPEN LIMITED, registered in England and Wales, registration number: 11086078, The Shard, 25th floor, 32 London Bridge Street
London, SE19SG – United Kingdom
Printed in Croatia

British Library Cataloguing-in-Publication Data
A catalogue record for this book is available from the British Library

Additional hard copies can be obtained from orders@intechopen.com

Optimization Algorithms - Examples, Edited by Jan Valdman
p. cm.
Print ISBN 978-1-78923-676-7
Online ISBN 978-1-78923-677-4

For EU product safety concerns:
IN TECH d.o.o., Prolaz Marije Krucifikse Kozulić 3, 51000 Rijeka, Croatia,
info@intechopen.com or visit our website at intechopen.com.

Meet the editor

Dr. Jan Valdman is an associate professor of applied mathematics at the University of South Bohemia in České Budějovice (Czech Republic) and a researcher at the Institute of Information Theory and Automation in Prague. He obtained his MSc degrees from the Mathematical Research Institute (Utrecht, the Netherlands) and the University of West Bohemia in Pilsen in 1998. He graduated from the University of Kiel (Germany) with his PhD thesis on modeling of elastoplasticity in 2002. After spending many years at several foreign institutions (Linz, Bergen, Reykjavik, and Leipzig), he returned to Czech Republic. He worked in applied mathematics at the Technical University Ostrava in 2011. His areas of interests include computational nonlinear mechanics of solids and a posteriori error estimates for nonlinear problems. He has published several vectorized codes in MATLAB for finite element assemblies.

Contents

Preface

The work on this book started in October 2017. At the beginning of 2018, I was asked by IntechOpen to become a new editor.

Since I had already served as an editor of the book *Applications from Engineering with MATLAB Concepts* in 2016, I accepted the offer. The general aim of this book is to present selected optimization algorithms in the areas of engineering, sciences and economics. During the selection process, many contributions were received, but some had to be rejected. These were mostly too theoretical and missing explanations of the algorithms. My intention was to provide a book with a focus on: (1) clear motivation of studied problems, (2) understanding of described algorithms by a broad spectrum of scientists, and (3) providing (computational) examples that a reader can easily repeat. At the final stage, only seven independent chapters remained. I hope our book entitled *Optimization Algorithms - Examples* will serve as a useful reference to students, scientists or engineers.

I am thankful to each author for the technical effort presented in each book chapter and their patience when working on revisions.

My biggest thanks go to Ms. Ivana Glavic, Author Service Manager from IntechOpen, who instructed me through many stages of the editorial process. Together, we did our best to ensure the book's high quality.

Dr. Jan Valdman
Institute of Mathematics and Biomathematics
University of South Bohemia
České Budějovice, Czech Republic

Institute of Information Theory and Automation of the ASCR
Prague, Czech Republic

Chapter 1

Distributionally Robust Optimization

Jian Gao, Yida Xu, Julian Barreiro-Gomez,
Massa Ndong, Michalis Smyrnakis and
Hamidou Tembine

Additional information is available at the end of the chapter

http://dx.doi.org/10.5772/intechopen.76686

Abstract

This chapter presents a class of distributionally robust optimization problems in which a decision-maker has to choose an action in an uncertain environment. The decision-maker has a continuous action space and aims to learn her optimal strategy. The true distribution of the uncertainty is unknown to the decision-maker. This chapter provides alternative ways to select a distribution based on empirical observations of the decision-maker. This leads to a distributionally robust optimization problem. Simple algorithms, whose dynamics are inspired from the gradient flows, are proposed to find local optima. The method is extended to a class of optimization problems with orthogonal constraints and coupled constraints over the simplex set and polytopes. The designed dynamics do not use the projection operator and are able to satisfy both upper- and lower-bound constraints. The convergence rate of the algorithm to generalized evolutionarily stable strategy is derived using a mean regret estimate. Illustrative examples are provided.

Keywords: distribution robustness, gradient flow, Bregman divergence, Wasserstein metric, f-divergence

1. Introduction

Robust optimization can be defined as the process of determining the best or most effective result, utilizing a quantitative measurement system under worst case uncertain functions or parameters. The optimization may occur in terms of best robust design, net cash flows, profits, costs, benefit/cost ratio, quality-of-experience, satisfaction, end-to-end delay, completion time, etc. Other measurement units may be used, such as units of production or production time, and optimization may occur in terms of maximizing production units, minimizing processing time, production time, maximizing profits, or minimizing costs under uncertain parameters. There are numerous techniques of robust optimization methods such as robust linear programming, robust dynamic programming, robust geometric programming, queuing theory, risk analysis, etc. One of the main drawbacks of the robust optimization is that the worst scenario may be too conservative. The bounds provided by the worst case scenarios may not be useful in many interesting problems (see the wireless communication example provided below). However, distributionally robust optimization is not based on the worst case parameters. The distributional robustness method is based the probability distribution instead of worst parameters. The worse case distribution within a certain carefully designed distributional uncertainty set may provide interesting features. Distributionally robust programming can be used not only to provide a distributionally robust solution to a problem when the true distribution is unknown, but it also can, in many instances, give a general solution taking into account some risk. The presented methodology is simple and reduces significantly the dimensionality of the distributionally robust optimization. We hope that the designs of distributionally robust programming presented here can help designers, engineers, cost–benefit analyst, managers to solve concrete problems under unknown distribution.

The rest of the chapter is organized as follows. Section 2 presents some preliminary concepts of distributionally robust optimization. A class of constrained distributionally robust optimization problems are presented in Section 3. Section 4 focuses on distributed distributionally robust optimization. Afterwards, illustrative examples in distributed power networks and in wireless communication networks are provided to evaluate the performance of the method. Finally, prior works and concluding remarks are drawn in Section 5.

Notation: Let $\mathbb{R}$, $\mathbb{R}_+$, denote the set of real and non-negative real numbers, respectively, (Ω, d) be a separable completely metrizable topological space with $d: \ \Omega \times \Omega \rightarrow \mathbb{R}_+$ a metric (distance). Let $\mathcal{P}(\Omega)$ be the set of all probability measures over Ω.

2. Distributionally robust optimization

This section introduces distributionally robust optimization models. We will first present a generic formulation of the problem. Then, individual components of the optimization and their solvability issues via equivalent formulations will be discussed.

2.1. Model

Consider a decision-maker who wants to select an action $a \in \mathcal{A} \subset \mathbb{R}^n$ in order to optimize her objective $r(a, \omega)$, where ω is an uncertain parameter. The information structure is the following:

- The true distribution of ω is not known to the decision-maker.
- The upper/lower bound (if any) of ω are unknown to the decision-maker.
- The decision-maker can measure/observe realization of the random variable ω.

The decision-maker chooses to experiment several trials and obtains statistical realizations of ω from measurements. The measurement data can be noisy, imperfect and erroneous. Then, an empirical distribution (or histogram) m is built from the realizations of ω. However, m is not the true distribution of the random variable ω, and m may not be a reliable measure due to statistical, bias, measurement, observation or computational errors. Therefore, the decision-maker is facing a risk. The risk-sensitive decision-maker should decide action that improves the performance of $\mathbb{E}_{\tilde{m}} r(a, \omega)$ among alternative distributions $\tilde{m}$ within a certain level of deviation $\rho > 0$ from the distribution m. The distributionally robust optimization problem is therefore formulated as

$$\sup_{a \in \mathcal{A}} \inf_{\tilde{m} \in B_\rho(m)} \mathbb{E}_{\omega \sim \tilde{m}} r(a, \omega). \tag{1}$$

where $B_\rho(m)$ is the uncertainty set of alternative admissible distributions from m within a certain radius $\rho > 0$. Different distributional uncertainty sets are presented: the f-divergence and the Wasserstein metric, defined below.

2.1.1. *f-divergence*

We introduce the notion of $f-$ divergence which will be used to compute the discrepancy between probability distributions.

Definition 1. *Let m and $\tilde{m}$ be two probability measures over Ω such that m is absolutely continuous with respect to $\tilde{m}$. Let f be a convex function. Then, the f-divergence between m and $\tilde{m}$ is defined as follows:*

$$D_f(m \| \tilde{m}) \equiv \int_\Omega f\left(\frac{dm}{d\tilde{m}}\right) d\tilde{m} - f(1),$$

where $\frac{dm}{d\tilde{m}}$ is the Radon-Nikodym derivative of the measure m with the respect the measure $\tilde{m}$.

By Jensen's inequality:

$$\begin{aligned} D_f(m \| \tilde{m}) &= \int_\Omega f\left(\frac{dm}{d\tilde{m}}\right) d\tilde{m} - f(1) \\ &\geq f\left(\int_\Omega \frac{dm}{d\tilde{m}} d\tilde{m}\right) - f(1) \\ &= f\left(\int_\Omega dm\right) - f(1) \\ &= f(1) - f(1) = 0. \end{aligned} \tag{2}$$

Thus, $D_f(m \| \tilde{m}) \geq 0$ for any convex function f. Note however that, the $f-$ divergence $D_f(m \| \tilde{m})$ is not a distance (for example, it does not satisfy the symmetry property). Here the distributional uncertainty set imposed to the alternative distribution $\tilde{m}$ is given by

$$B_\rho(m) = \left\{ \tilde{m} | \tilde{m}(.) \geq 0, \ \int_\Omega d\tilde{m} = \tilde{m}(\Omega) = 1, \ D_f(\tilde{m} \| m) \leq \rho \right\}.$$

Example 1. *From the notion of f– divergence one can derive the following important concept:*

- *α-divergence for*

$$f(a) = \begin{cases} \frac{4}{(\alpha+1)(1-\alpha)}\left(1 - a^{\frac{\alpha+1}{2}}\right) & \text{if } \alpha \notin \{-1, +1\}, \\ a \log a & \text{if } \alpha = 1, \\ -\log a & \text{if } \alpha = -1, \end{cases}$$

- *In particular, Kullback–Leibler divergence (or relative entropy) is retrieved as α goes to* 1.

2.1.2. Wasserstein metric

The Wasserstein metric between two probability distributions $\tilde{m}$ and m is defined as follows:

Definition 2. *For $m, \tilde{m} \in \mathcal{P}(\Omega)$, let $\Pi(\tilde{m}; m)$ be the set of all couplings between m and $\tilde{m}$. That is,*

$$\{\pi \in \mathcal{P}(\Omega \times \Omega) \mid \pi(A \times \Omega) = m(A), \ \pi(\Omega \times B) = \tilde{m}(B), \ (A, B) \in \mathcal{B}^2(\Omega)\}.$$

$\mathcal{B}(\Omega)$ denotes the measurable sets of Ω. Let $\theta \in [1, \infty]$. The Wasserstein metric between $\tilde{m}$ and m is defined as

$$W_\theta(\tilde{m}; m) = \inf_{\pi \in \Pi(\tilde{m}; m)} \|d\|_{L^\theta_\pi} = \inf_{\pi \in \Pi(\tilde{m}; m)} \int_{(a,b)} d^\theta(a, b)\pi(da, db),$$

It is well-known that for every $\theta \geq 1$, $W_\theta(\tilde{m}; m)$ is a true distance in the sense that it satisfies the following three axioms:

- positive-definiteness,
- the symmetry property,
- the triangle inequality.

Note that $\tilde{m}$ is not necessarily absolutely continuous with respect to m. Now the distributional uncertainty/constraint set is the set of all possible probability distributions within a L^θ-Wasserstein distance below ρ.

$$\tilde{B}_\rho(m) = \left\{ \tilde{m} | \int_\Omega d\tilde{m} = \tilde{m}(\Omega) = 1, \ W_\theta(\tilde{m}; m) \leq \rho \right\},$$

Note that, if m is a random measure (obtained from a sampled realization), we use the expected value of the Wasserstein metric.

Example 2. *The L^{θ}-Wasserstein distance between two Dirac measures δ_{ω_0} and $\delta_{\tilde{\omega}_0}$ is $W_\theta(\delta_{\omega_0}, \delta_{\tilde{\omega}_0}) = d(\omega_0, \tilde{\omega}_o)$. More generally, for $K \geq 2$, the L^2-Wasserstein distance between empirical measures $\mu_K = \frac{1}{K}\sum_{k=1}^{K} \delta_{\omega_k}$ and $\nu_K = \frac{1}{K}\sum_{k=1}^{K} \delta_{\tilde{\omega}_k}$ is $W_2^2(\mu_K, \nu_K) \leq \frac{1}{K}\sum_{i=1}^{K} [\omega_k - \tilde{\omega}_k]^2$.*

We have defined $B_\rho(m)$ and $\tilde{B}_\rho(m)$. The goal now is to solve (1) under both $f-$ divergence and Wasserstein metric. One of the difficulties of the problem is the curse of dimensionality. The distributionally robust optimization problem (1) of the decision-maker is an infinite-dimensional robust optimization problem because B_ρ is of infinite dimensions. Below we will show that (1) can be transformed into an optimization in the form of supinfsup. The latter problem has three alternating terms. Solving this problem requires a triality theory.

2.2. Triality theory

We first present the duality gap and develop a triality theory to solve equivalent formulations of (1). Consider uncoupled domains $\mathcal{A}_i, i \in \{1, 2, 3\}$. For a general function r_2, one has

$$\sup_{a_2 \in \mathcal{A}_2} \inf_{a_1 \in \mathcal{A}_1} r_2(a_1, a_2) \leq \inf_{a_1 \in \mathcal{A}_1} \sup_{a_2 \in \mathcal{A}_2} r_2(a_1, a_2)$$

and the difference

$$\min_{a_1 \in \mathcal{A}_1} \max_{a_2 \in \mathcal{A}_2} r_2(a_1, a_2) - \max_{a_2 \in \mathcal{A}_2} \min_{a_1 \in \mathcal{A}_1} r_2(a_1, a_2),$$

is called duality gap. As it is widely known in duality theory from Sion's Theorem [1] (which is an extension of von Neumann minimax Theorem) the duality gap vanishes, for example for convex-concave function, and the value is achieved by a saddle point in the case of non-empty convex compact domain.

Triality theory focuses on optimization problems of the forms: sup infsup or infsup inf. The term triality is used here because there are three key alternating terms in these optimizations.

Proposition 1. *Let $(a_1, a_2, a_3) \mapsto r_3(a_1, a_2, a_3) \in \mathbb{R}$ be a function defined on the product space $\prod_{i=1}^{3} \mathcal{A}_i$. Then, the following inequalities hold:*

$$\begin{aligned} &\sup_{a_2 \in \mathcal{A}_2} \inf_{a_1 \in \mathcal{A}_1, a_3 \in \mathcal{A}_3} r_3(a_1, a_2, a_3) \leq \\ &\inf_{a_3 \in \mathcal{A}_3} \sup_{a_2 \in \mathcal{A}_2} \inf_{a_1 \in \mathcal{A}_1} r_3(a_1, a_2, a_3) \leq \\ &\inf_{a_1 \in \mathcal{A}_1, a_3 \in \mathcal{A}_3} \sup_{a_2 \in \mathcal{A}_2} r_3(a_1, a_2, a_3), \end{aligned} \tag{3}$$

and similarly

$$\begin{aligned} &\sup_{a_1 \in \mathcal{A}_1, a_3 \in \mathcal{A}_3} \inf_{a_2 \in \mathcal{A}_2} r_3(a_1, a_2, a_3) \leq \\ &\sup_{a_3 \in \mathcal{A}_3} \inf_{a_2 \in \mathcal{A}_2} \sup_{a_1 \in \mathcal{A}_1} r_3(a_1, a_2, a_3) \leq \\ &\inf_{a_2 \in \mathcal{A}_2} \sup_{a_1 \in \mathcal{A}_1, a_3 \in \mathcal{A}_3} r_3(a_1, a_2, a_3). \end{aligned} \tag{4}$$

Proof. Define

$$\hat{g}(a_2, a_3) := \inf_{a_1 \in \mathcal{A}_1} r_3(a_1, a_2, a_3).$$

Thus, for all a_2, a_3, one has $\hat{g}(a_2, a_3) \leq r_3(a_1, a_2, a_3)$. It follows that, for any a_1, a_3,

$$\sup_{a_2 \in \mathcal{A}_2} \hat{g}(a_2, a_3) \leq \sup_{a_2 \in \mathcal{A}_2} r_3(a_1, a_2, a_3).$$

Using the definition of $\hat{g}$, one obtains

$$\sup_{a_2 \in \mathcal{A}_2} \inf_{a_1 \in \mathcal{A}_1} r_3(a_1, a_2, a_3) \leq \sup_{a_2 \in \mathcal{A}_2} r_3(a_1, a_2, a_3), \ \forall a_1, a_3.$$

Taking the infimum in a_1 yields:

$$\sup_{a_2 \in \mathcal{A}_2} \inf_{a_1 \in \mathcal{A}_1} r_3(a_1, a_2, a_3) \leq \inf_{a_1 \in \mathcal{A}_1} \sup_{a_2 \in \mathcal{A}_2} r_3(a_1, a_2, a_3), \ \forall a_3 \tag{5}$$

Now, we use two operations for the variable a_3:

- Taking the infimum in the inequality (5) in a_3 yields

$$\begin{aligned} \inf_{a_3 \in \mathcal{A}_3} \sup_{a_2 \in \mathcal{A}_2} \inf_{a_1 \in \mathcal{A}_1} r_3(a_1, a_2, a_3) &\leq \inf_{a_3 \in \mathcal{A}_3} \inf_{a_1 \in \mathcal{A}_1} \sup_{a_2 \in \mathcal{A}_2} r_3(a_1, a_2, a_3) \\ &= \inf_{(a_1, a_3) \in \mathcal{A}_1 \times \mathcal{A}_3} \sup_{a_2 \in \mathcal{A}_2} r_3(a_1, a_2, a_3), \end{aligned}$$

which proves the second part of the inequalities (3). The first part of the inequalities (3) follows immediately from (5).

- Taking the supremum in inequality (5) in a_3 yields

$$\sup_{(a_2, a_3) \in \mathcal{A}_2 \times \mathcal{A}_3} \inf_{a_1 \in \mathcal{A}_1} r_3(a_1, a_2, a_3) \leq \sup_{a_3 \in \mathcal{A}_3} \inf_{a_1 \in \mathcal{A}_1} \sup_{a_2 \in \mathcal{A}_2} r_3(a_1, a_2, a_3),$$

which proves the first part of the inequalities (4). The second part of the inequalities (4) follows immediately from (5).

This completes the proof.

2.3. Equivalent formulations

Below we explain how the dimensionality of problem (1) can be significantly reduced using a representation by means of the triality theory inequalities of Proposition 1.

2.3.1. *f-divergence*

Interestingly, the distributionally robust optimization problem (1) under f-divergence is equivalent to the finite dimensional stochastic optimization problem (when $\mathcal{A}$ are of finite dimensions). To see this, the original problem need to be transformed. Let us introduce the likelihood functional $L(\tilde{\omega}) = \frac{d\tilde{m}}{dm}(\tilde{\omega})$, and set

$$L_\rho(m) = \left\{ L | \int_{\tilde{\omega}} f(L(\tilde{\omega}))dm - f(1) \leq \rho, \quad \int_{\tilde{\omega}} L(\tilde{\omega})dm(\tilde{\omega}) = 1 \right\}.$$

Then, the Lagrangian of the problem is

$$\begin{aligned}\tilde{r}(a, L, \lambda, \mu) = & \int_{\tilde{\omega}} r(a, \tilde{\omega})L(\tilde{\omega})dm(\tilde{\omega}) \\ & - \lambda\left(\rho + f(1) - \int_{\tilde{\omega}} f(L(\tilde{\omega}))dm(\tilde{\omega})\right) \\ & - \mu\left(1 - \int_{\tilde{\omega}} L(\tilde{\omega})dm(\tilde{\omega})\right),\end{aligned}$$

where $\lambda \geq 0$ and $\mu \in \mathbb{R}$. The problem becomes

$$\left\{ \sup_a \inf_{L \in L_\rho(m)} \sup_{\lambda \geq 0, \mu \in \mathbb{R}} \tilde{r}(a, L, \lambda, \mu). \right. \tag{6}$$

A full understanding of problem (6) requires a triality theory (not a duality theory). The use of triality theory leads to the following equation:

$$\left\{ \sup_{a \in \mathcal{A}} \inf_{\tilde{m} \in B_\rho(m)} \mathbb{E}_{\tilde{m}}[r] = \sup_{a \in \mathcal{A}, \lambda \geq 0, \mu \in \mathbb{R}} \mathbb{E}_m h, \right. \tag{7}$$

where h is the integrand function $-\lambda(\rho + f(1)) - \mu - \lambda f^*\left(\frac{r+\mu}{-\lambda}\right)$, where f^* is Legendre-Fenchel transform of f defined by

$$f^*(\xi) = \sup_L [\langle L, \xi \rangle - f(L)] = -\inf_L [f(L) - \langle L, \xi \rangle]. \tag{8}$$

Note that the righthand side of (7) is of dimension $n + 2$, which reduces considerably the dimensionality of the original problem (1).

2.3.2. *Wasserstein metric*

Similarly, the distributionally robust optimization problem under Wasserstein metric is equivalent to the finite dimensional stochastic optimization problem (when $\mathcal{A}$ is a set of finite dimension). If the function $\omega \mapsto r(a, \omega)$ is upper semi-continuous and (Ω, d) is a Polish space then the Wasserstein distributionally robust optimization problem is equivalent to

$$\begin{cases} \sup_{a \in \mathcal{A}} \inf_{\tilde{m} \in \tilde{B}_\rho}(m) \mathbb{E}_{\tilde{m}}[r] = \sup_{a \in \mathcal{A}} \sup_{\lambda \geq 0} \mathbb{E}_m\left[\tilde{h}\right], \\ \tilde{h} = \lambda\rho^\theta + \mu + \sup_{\hat{\omega} \in \Omega}\left[r(a, \omega) - \mu - \lambda d^\theta(\omega, \hat{\omega})\right]; \end{cases} \tag{9}$$

The next subsection presents algorithms for computing a distributionally robust solution from the equivalent formulations above.

2.4. Learning algorithms

Learning algorithms are crucial for finding approximate solutions to optimization and control problems. They are widely used for seeking roots/kernel of a function and for finding feasible solutions to variational inequalities. Practically, a learning algorithm generates a certain trajectory (or a set of trajectories) toward a potential approximate solution. Selecting a learning algorithm that has specific properties such as better accuracy, more stability, less-oscillatory and quick convergence is a challenging task [2–5]. From the calculus of variations point of view, however, a learning algorithm generates curves. Therefore, selecting an algorithm among the others leads to an optimal control problem on the spaces of curves. Hence, it is natural to use optimal control theory to derive faster algorithms for a family of curves. Bergman-based algorithms and risk-aware version of it are introduced below to meet specific properties. We start by introducing the Bregman divergence.

Definition 3. *The Bregman divergence $d_g : \mathcal{A} \times \mathcal{A} \to \mathbb{R}$ is defined on a differentiable strictly convex function $g : \mathcal{A} \to \mathbb{R}$. For two points $(a, b) \in \mathcal{A}^2$, it measures the gap between $g(a)$ and the first-order Taylor expansion of g around a evaluated at b*

$$d_g(a, b) := g(a) - g(b) - \langle \nabla g(b), a - b \rangle.$$

Example 3. *From the Bregman divergence one gets other features by choosing specific functions g :*

- *If $g(a) = \sum_{i=1}^{n} a_i^2$ then the Bregman divergence $d_g(a, b) = \sum_{i=1}^{n} (a_i - b_i)^2$ is the squared standard Euclidean distance.*
- If $g(a) = \sum_{i=1}^{n} a_i \log a_i$ is defined on the relative interior of the simplex, i.e., $a \in \{b \mid b \in (0,1)^n,\ \sum_{i=1}^{n} b_i = 1\}$ then the Bregman divergence $d_g(a, b) = \sum_{i=1}^{n} a_i \log\left(\frac{a_i}{b_i}\right)$, is the Kullback–Leibler divergence.

We are now ready to define algorithms for solving the righthand side of (7) and (9). One of the key approaches for error quantification of the algorithm with respect to the distributionally robust optimum is the so-called average regret. When the regret vanishes one gets close to a distributionally robust optimum.

Definition 4. *The average regret of an algorithm which generates the trajectory $a(t) = (\tilde{a}(t), \lambda(t), \mu(t))$ within $[t_0, T]$, $t_0 > 0$ is*

$$\text{regret}_T := \frac{1}{T - t_0} \int_{t_0}^{T} \left[\max_{b \in \mathcal{A} \times \mathbb{R}_+ \times \mathbb{R}} \mathbb{E}_m h(b, \omega) \right] - \mathbb{E}_m h(a(t), \omega) dt$$

2.4.1. *Armijo gradient flow*

Algorithm 1. *The Armijo's gradient pseudocode is as follows:*

1: ***Procedure*** ARMIJO GRADIENT $(a(0), \epsilon, T, g, m, h)$ ▷ *The Armijo's gradient starting from $a(0)$ within* $[0, T]$

http://dx.doi.org/10.5772/intechopen.76686

```
    a ← a(0)
    while regret > ϵ and t ≤ T do ▷ We have the answer if regret is 0
        Compute a(t) solution of (10)
        Compute regret_t
    end while
    return a(t), regret_t ▷ get a(t) and the regret
end procedure
```

Proposition 2. *Let* $a \mapsto \mathbb{E}_m h(a,\omega) : \mathbb{R}^{n+2} \to \mathbb{R}$ *be a concave function that has a unique global maximizer* a^*. *Assume that* a^* *be a feasible action profile, i.e.,* $a^* \in \mathcal{A}$. *Consider the continuous time analogue of the Armijo gradient flow* [6], *which is given by*

$$\begin{aligned}&\frac{d}{dt}a(t) = \left[\nabla^2 g\right]^{-1}.\nabla_a \mathbb{E}_m h(a(t),\omega),\\&a(0) = a_0 \in \mathbb{R}^{n+2},\end{aligned} \tag{10}$$

where $a(0) = a_0$ is the initial point of the algorithm and g is a strictly convex function on a. Let $a(t)$ be the solution to (10).

Then the average regret within $[t_0, T]$, $t_0 > 0$ *is bounded above by*

$$regret_T := \frac{1}{T-t_0}\int_{t_0}^{T} \mathbb{E}_m[h(a^*,\omega) - h(a(t),\omega)]dt \leq d_g(a^*, a_0)\frac{\log\frac{T}{t_0}}{T-t_0}.$$

Proof. Let

$$W(a(t)) = t\mathbb{E}_m[h(a^*,\omega) - h(a(t),\omega)] + d_g(a^*, a(t)),$$

where a is solution to (10). The function W is positive and $\frac{d}{dt}W = \mathbb{E}_m[h(a^*,\omega) - h(a(t),\omega)] - t\left\langle \mathbb{E}_m \nabla_a h(a,\omega), g_{aa}^{-1}\mathbb{E}_m \nabla_a h(a(t),\omega)\right\rangle + \frac{d}{dt}d_g(a^*, a(t))$. By concavity of $\mathbb{E}_m h(a,\omega)$ one has

$$\langle \mathbb{E}_m \nabla_a h(a,\omega), (a^* - a)\rangle \geq \mathbb{E}_m[h(a^*,\omega) - h(a,\omega)], \ \ \forall\ a.$$

On the other hand,

$$\begin{aligned}&\frac{d}{dt}d_g(a^*, a(t)) = -\dot{a}g_a(a) - \left\langle g_{aa}\dot{a}, a - a^*\right\rangle + g_a\dot{a}\\&= -\left\langle g_{aa}\dot{a}, a - a^*\right\rangle = -\langle \mathbb{E}_m \nabla_a h(a,\omega), a^* - a\rangle.\end{aligned} \tag{11}$$

Hence,

$$\begin{aligned}&\frac{d}{dt}W \leq \langle \mathbb{E}_m \nabla_a h(a,\omega), (a^* - a)\rangle\\&-t\left\langle \mathbb{E}_m \nabla_a h(a,\omega), g_{aa}^{-1}\mathbb{E}_m \nabla_a h(a,\omega)\right\rangle\\&-\langle \mathbb{E}_m \nabla_a h(a,\omega), a^* - a\rangle\\&= -t\left\langle \mathbb{E}_m \nabla_a h(a,\omega), g_{aa}^{-1}\mathbb{E}_m \nabla_a h(a,\omega)\right\rangle \leq 0,\end{aligned} \tag{12}$$

where the last inequality is by convexity of g. It follows that $\frac{d}{dt}W(a(t)) \leq 0$ along the path of the gradient flow. This decreasing property implies $0 \leq W(a(t)) \leq W(a(0)) = d_g(a^*, a(0))$. In particular, $0 \leq t\mathbb{E}_m[h(a^*, \omega) - h(a, \omega)] \leq W(a(0)) < +\infty$. Thus, the error to the value $\mathbb{E}_m h(a^*, \omega)$ is bounded by

$$0 \leq \mathbb{E}_m[h(a^*, \omega) - h(a, \omega)] \leq \frac{W(a(0))}{t}.$$

The announced result on the regret follows by integration over $[t_0, T]$ and by averaging. This completes the proof.

Note that the above regret-bound is established without assuming strong convexity of $a \mapsto -\mathbb{E}_m h(a, \omega)$. Also no Lipschitz continuity bound of the gradient is assumed.

2.4.2. Bregman learning algorithms

Algorithm 2. *The Bregman learning pseudocode is as follows:*

procedure BREGMAN $(a(0), \epsilon, T, g, \alpha, \beta, m, h)$ ▷ *The Bregman learning starting from $a(0)$ within $[0, T]$*

  $a \leftarrow a(0)$

  while *regret* $> \epsilon$ *and* $t \leq T$ ***do*** ▷ *We have the answer if regret is 0*

    Compute $a(t)$ solution of (13)

    Compute regret$_t$

  end while

  return $a(t)$, *regret$_t$* ▷ *get $a(t)$ and the regret*

end procedure

Proposition 3. *Let $a \mapsto \mathbb{E}_m h(a, \omega) : \mathbb{R}^{n+2} \to \mathbb{R}$ be a concave function that has a unique global maximizer a^*. Assume that a^* be a feasible action profile, i.e., $a^* \in \mathcal{A}$. Let α and β be two functions such that $\dot{\beta}(t) \leq e^{\alpha(t)}$. Consider the following Bregman learning algorithm*

$$\begin{aligned} &\frac{d}{dt}\left[g_a\left(a(t) + e^{-\alpha(t)}\dot{a}(t)\right)\right] = e^{\alpha(t)+\beta(t)}\nabla_a \mathbb{E}_m h(a(t), \omega), \\ &a(0) \in \mathbb{R}^{n+2}, \dot{a}(0) \in \mathbb{R}^{n+2}, \end{aligned} \tag{13}$$

where $a(0)$ is the initial point of the algorithm and g is a strictly convex function on a. Let $a(t)$ be the solution to (13). Then the average regret within $[t_0, T]$, $t_0 > 0$ is bounded above by

$$regret_T \leq \frac{c_0}{T - t_0}\int_{t_0}^{T} e^{-\beta(s)} ds, \tag{14}$$

where $c_0 := d_g(a^*, a(0)) + e^{-\alpha(0)}\dot{a}(0)) + e^{\beta(0)}\mathbb{E}_m[h(a^*, \omega) - h(a(0), \omega)] > 0.$

Proof. Let $W(a, \dot{a}, t, a^*) = d_g\left(a^*, a(t) + e^{-\alpha(t)}\dot{a}(t)\right) + e^{\beta(t)}\mathbb{E}_m[h(a^*, \omega) - h(a(t), \omega)]$. It is clear that W is positive. Moreover, $\frac{d}{dt}W(a(t), \dot{a}(t), t, a^*) \leq 0$ for $\dot{\beta} \leq e^{\alpha}$. Thus $W(a(t), \dot{a}(t), t, a^*) \leq W(a(0), \dot{a}(0), 0, a^*) = c_0$. By integration between $[t_0, T]$ it follows

$$\frac{1}{T - t_0}\int_{t_0}^{T} \mathbb{E}_m[h(a^*, \omega) - h(a(t), \omega)]dt \leq \frac{c_0}{T - t_0}\int_{t_0}^{T} e^{-\beta(s)}ds.$$

This completes the proof.

In particular, for $\beta(s) = -s + e^s$, one obtains an error bound to the minimum value as

$$\frac{c_0}{t}\int_0^t e^{-\beta(s)}ds = \frac{c_0}{t}\int_0^t e^s e^{-e^s}ds = \frac{c_0\left(\frac{1}{e} - e^{-e^t}\right)}{t},$$

and for $\beta(s) = s$, the regret bound becomes

$$\frac{c_0}{t}\int_0^t e^{-\beta(s)}ds = \frac{c_0(1 - e^{-t})}{t}.$$

Figure 1 illustrates the advantage of algorithm (13) compared with the gradient flow (10). It plots the regret bound $\frac{c_0}{T-t_0}\int_{t_0}^{T} e^{-\beta(s)}ds$ for $\beta = s$ and $d_g(a^*, a_0)\frac{\log\frac{T}{t_0}}{T-t_0}$ with an initial gap of $c_0 = 25$.

The advantage of algorithms (10) and (13) is that it is not required to compute the Hessian of $\mathbb{E}_m h(a, \omega)$ as it is the case in the Newton scheme. As a corollary of Proposition 2 the regret vanishes as T grows. Thus, it is a no-regret algorithm. However, Algorithm (10) may not be sufficiently fast. Algorithm (13) provides a higher order convergence rate by carefully designing (α, β). The average regret decays very quickly to zero [7]. However, it may generate an

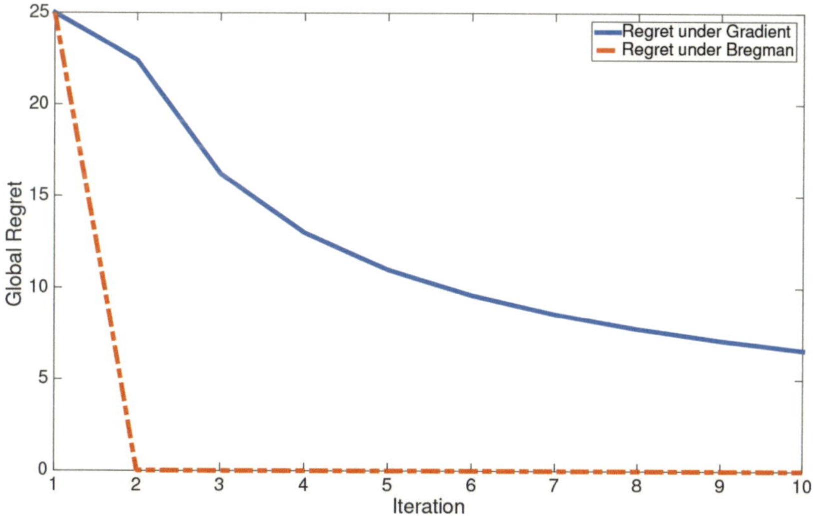

Figure 1. Global regret bound under Bregman vs. gradient. The initial gap is $c_0 = 25$.

oscillatory trajectory with a big magnitude. The next subsection presents risk-aware algorithms that reduce the oscillatory phase of the trajectory.

2.4.3. Risk-aware Bregman learning algorithm

In order to reduce the oscillatory phase, we introduce a risk-aware Bregman learning algorithm [7] which is a speed-up-and-average version of (13) called *mean dynamics* $\overline{m}$ of a given by

$$\begin{aligned}\dddot{\overline{m}} &= -\frac{3}{t}\ddot{\overline{m}} - (e^{\alpha} - \dot{\alpha})\left(\ddot{\overline{m}} + \frac{2}{t}\dot{\overline{m}}\right) \\ &\quad + \frac{e^{2\alpha+\beta}}{t} g^{-1}_{\overline{m}\,\overline{m}}(\overline{m} + [t + 2e^{-\alpha}]\dot{\overline{m}} + te^{-\alpha}\ddot{\overline{m}})\mathbb{E}h_{\overline{m}}(t\dot{\overline{m}} + \overline{m}, \omega),\end{aligned} \tag{15}$$

with starting vector $\overline{m}(0) = a(0), \dot{\overline{m}}(0), \ddot{\overline{m}}(0)$.

Algorithm 3. *The risk-aware Bregman learning pseudocode is as follows:*

1: ***procedure*** RISK-AWARE BREGMAN $(\overline{m}(0), \epsilon, T, g, \alpha, \beta, m, h)$ ▷ *The risk-aware Bregman learning starting from* $\overline{m}(0)$ *within* $[0, T]$

2: $\overline{m} \leftarrow \overline{m}(0) = a(0), \dot{\overline{m}}(0), \ddot{\overline{m}}(0)$

3: ***while*** *regret* $> \epsilon$ *and* $t \leq T$ ***do*** ▷ *We have the answer if regret is 0*

4: *Compute* $\overline{m}(t)$ *solution of (15)*

5: *Compute regret*

6: ***end while***

7: ***return*** $\overline{m}(t)$, $regret_t$ ▷ *get* $\overline{m}(t)$ *and the regret*

8: ***end procedure***

Proposition 4. *The time-average trajectory of the learning algorithm (13) generates the mean dynamics (15).*

Proof. We use the average relation $\overline{m}(t) = \frac{1}{t}\int_0^t a(s)\ ds$ where a solves Eq. (13). From the definition of $\overline{m}$, and by Hopital's rule, $\overline{m}(0) = a(0)$. Moreover, $\overline{m}(t)$ and $a(t)$ share the following equations:

$$\begin{aligned} a(t) &= \overline{m}(t) + t\dot{\overline{m}}(t), \\ \dot{a}(t) &= 2\dot{\overline{m}}(t) + t\ddot{\overline{m}}(t), \\ \ddot{a}(t) &= 3\ddot{\overline{m}}(t) + t\dddot{\overline{m}}(t). \end{aligned} \tag{16}$$

Substituting these values in Eq. (13) yields the mean dynamics (15). This completes the proof.

The risk-aware Bregman dynamics (15) generates a less oscillatory trajectory due to its averaging nature. The next result provides an accuracy bound for (15).

Proposition 5. *The risk-aware Bregman dynamics (15) satisfies*

$$0 \leq \mathbb{E}_m[h(a^*,\omega) - h(\overline{m}(t),\omega)] \leq \frac{c_0}{t}\int_0^t e^{-\beta(s)}ds.$$

Proof. Let $\overline{m}(t) = \frac{1}{t}\int_0^t a(s)ds$. Then, $\overline{m}(t) = \int_{\mathbb{R}} a(s)\left(\frac{1}{t}1\!\!1_{[0,t]}(s)\right)\ ds$. Thus, $\overline{m}(t) = \mathbb{E}_{\mu(t)}a$ where $\mu(t)$ is the measure with density $d\mu(t)[s] = \frac{1}{t}1\!\!1_{[0,t]}(ds)$. By convexity of $-\mathbb{E}_m h(a,\omega)$ we apply the Jensen's inequality:

$$\mathbb{E}_m h\left(\frac{1}{t}\int_0^t a(s)ds, \omega\right) = \mathbb{E}_m h(\overline{m}(t),\omega) = \mathbb{E}_m h\left(\mathbb{E}_{\mu(t)}a,\omega\right)$$

$$\geq \mathbb{E}_{\mu(t)}\mathbb{E}_m h(a,\omega) = \frac{1}{t}\int_0^t \mathbb{E}_m h(a(s),\omega)ds.$$

In view of (14) one has

$$0 \leq \mathbb{E}_m h(a^*,\omega) - \mathbb{E}_m h\left(\frac{1}{t}\int_0^t a(s)ds, \omega\right)$$

$$\leq \frac{1}{t}\int_0^t [\mathbb{E}_m h(a^*,\omega) - \mathbb{E}_m h(a(s),\omega)]ds$$

$$\leq c_0 \frac{1}{t}\int_0^t e^{-\beta(s)}ds,$$

$$0 \leq \mathbb{E}_m h(a^*,\omega) - \mathbb{E}_m h(\overline{m}(t),\omega) \leq \frac{c_0}{t}\int_0^t e^{-\beta(s)}ds.$$

This completes the proof.

Definition 5. (Convergence time). *Let $\delta > 0$ and $a(t)$ be the trajectory generated by Bregman algorithm starting from a_0 at time t_0. The convergence time to be within a ball $B(\mathbb{E}_m h(a^*,\omega),\delta)$ of radius $\delta > 0$ from the center $r(a^*)$ is given by*

$$T_\delta = \inf\{t \mid \mathbb{E}_m[h(a^*,\omega) - h(a(t),\omega)] \leq \delta,\ \ t > t_0\}.$$

Proposition 6. *Under the assumptions above, the error generated by the algorithm is at most (14) which means that it takes at most $T_\delta = \beta^{-1}\left[\log\frac{c_0}{\delta}\right]$ time units to the algorithm to be within a ball $B(r(a^*),\delta)$ of radius $\delta > 0$ from the center $\mathbb{E}_m h(a^*,\omega)$.*

Proof. The proof is immediate. For $\delta > 0$ the average regret bound of Proposition 5,

Convergence	Error bound	Time-to-reach T_δ
Triple exponential	$e^{-e^{e^t}} c_0$	$\log\left[\log\left(\log \frac{c_0}{\delta}\right)\right]$
$\alpha(t) = t + e^t,\ \ \beta(t) = e^{e^t}$		
Double exponential rate	$e^{-e^t} c_0$	$\log\left(\log \frac{c_0}{\delta}\right)$
$\alpha(t) = t,\ \ \beta(t) = e^t$		
Exponential rate	$e^{-t} c_0$	$\log \frac{c_0}{\delta}$
$\alpha(t) = 0,\ \ \beta(t) = t$		
Polynomial order k	$\frac{c_0}{t^k}$	$\frac{c_0^{1/k}}{\delta^{1/k}}$
$\alpha(t) = \log k - \log t,\ \ \beta(t) = k \log t$		

Table 1. Convergence rate under different set of functions.

$$regret_T \leq \frac{c_0}{T - t_0} \int_{t_0}^{T} e^{-\beta(s)} ds \leq \delta, \tag{17}$$

provides the announced convergence time bound. This completes the proof.

See **Table 1** for detailed parametric functions on the bound T_δ.

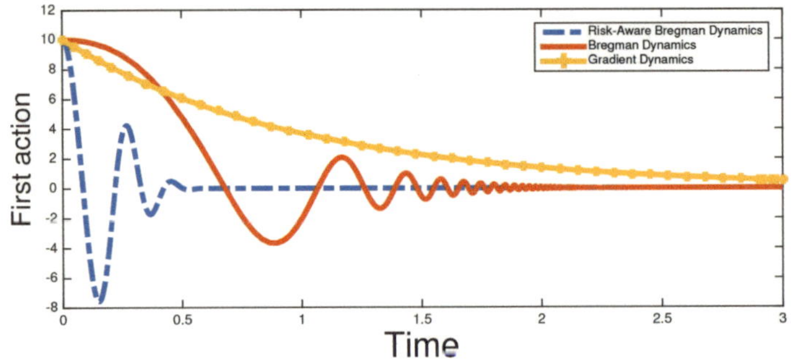

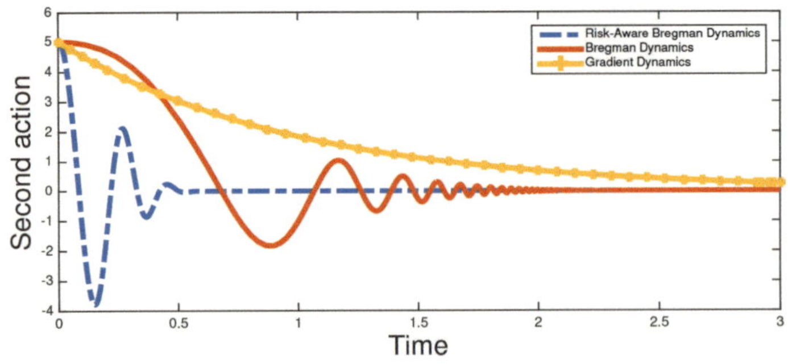

Figure 2. Gradient ascent vs. risk-aware Bregman dynamics for $r = -\left(1 + \sum_{k=1}^{2} \omega_k^2 a_k^2\right)$.

Example 4. *Let* $f(y) = y\log y$ *defined on* $\mathbb{R}_+^*$. *Then,* $f(1) = 0$, *and derivatives of* f *are* $f'(y) = 1 + \log y$, $f''(y) = \frac{1}{y} > 0$. *The Legendre-Fenchel transform of* f *is* $f^*(\xi) = y^* = e^{\xi-1}$. *Let* $(a_1, a_2) \mapsto g(a) = \|a\|_2^2$, *and* $(a_1, a_2, \omega) \mapsto r(a_1, a_2, \omega) = -\left(1 + \sum_{k=1}^2 \omega_{ik}^2 a_k^2\right)$. *The coefficient* ω *distribution is unknown but a sampled empirical measure* m *is considered to be similar to uniform distribution in* $(0, 1]$ *with* 10^4 *samples. We illustrate the quick convergence rate of the algorithm in a basic example and plot in* ***Figure 2*** *the trajectories under standard gradient, Bregman dynamics and risk-aware Bregman dynamics (15). In particular, we observe that risk-aware Bregman dynamics (15) provides very quickly a satisfactory value. In this particular setup, we observe that the accuracy of the risk-aware Bregman algorithm (15) at* $t = 0.5$ *will need four times* ($t = 2$) *less than the standard Bregman algorithm to reach a similar level of error. It takes* 40 *times more* ($t = 20$) *than the gradient ascent to reach that level. Also, we observe that the risk-aware Bregman algorithm is less oscillatory and the amplitude decays very fast compared to the risk-neutral algorithm.*

3. Constrained distributionally robust optimization

In the constrained case i.e., when $\mathcal{A}$ is a strict subset of $\mathbb{R}^{n+2}$, algorithms (10) and (13) present some drawbacks: The trajectory $a(t)$ may not be feasible, i.e., $a(t) \notin \mathcal{A} \times \mathbb{R}_+ \times \mathbb{R}$ even when it starts in $\mathcal{A}$. In order to design feasible trajectories, projected gradient has been widely studied in the literature. However, a projection into $\mathcal{A}$ at each time t involves additional optimization problems and the computation of the projected gradient adds extra complexity to the algorithm. We restrict our attention to the following constraints:

$$\mathcal{A} = \left\{ a \in \mathbb{R}^n \mid a_l \in \left[\underline{a}_l, \bar{a}_l\right],\ l \in \{1, \ldots, n\},\ \sum_{l=1}^n c_l a_l \leq b \right\}.$$

We impose the following feasibility condition: $\underline{a}_l < \bar{a}_l$, $l \in \{1, \ldots, n\}, c_l > 0$, $\sum_{l=1}^n c_l \underline{a}_l < b$. Under this setting, the constraint set $\mathcal{A}$ is non-empty, convex and compact.

We propose a method to compute a constrained solution that has a full support (whenever it exists). We do not use the projection operator. Indeed we transform the domain $\left[\underline{a}_l, \bar{a}_l\right] = \xi([0, 1])$ where $\xi(x_l) = \bar{a}_l x_l + \underline{a}_l(1 - x_l) = a_l$. ξ is a one-to-one mapping and

$$x_l = \xi^{-1}(a_l) = \frac{a_l - \underline{a}_l}{\bar{a}_l - \underline{a}_l} \in [0, 1].$$

$$\sum_{l=1}^n c_l \left(\bar{a}_l - \underline{a}_l\right) x_l \leq b - \sum_{l=1}^n c_l \underline{a}_l =: \hat{b}.$$

The algorithm (18)

$$\begin{cases} \dot{y} = [\nabla^2 g]^{-1} \nabla_a \mathbb{E}_m h(a, \omega) =: \hat{f}(a), \\ a_l := \bar{a}_l x_l + \underline{a}_l (1 - x_l), \\ x_l = \min\left(1, \frac{e^{y_l}}{\sum_{k=1}^{n} e^{y_k}} \frac{\hat{b}}{[c_l(\bar{a}_l - \underline{a}_l)]}\right), \\ l \in \{1, \ldots, n\}, \end{cases} \tag{18}$$

generates a trajectory $a(t)$ that satisfies the constraint.

Algorithm 4. *The constrained learning pseudocode is as follows:*

1: ***procedure*** CONSTRAINED GRADIENT $(a(0), \epsilon, T, g, m, h)$ ▷ *The constrained learning algorithm starting from $a(0)$ within $[0, T]$*

2: $a \leftarrow a(0)$

3: ***while*** *regret* $> \epsilon$ *and* $t \leq T$ ***do*** ▷ *We have the answer if regret is 0*

4: *Compute $a(t)$ solution of (18)*

5: *Compute regret*

6: ***end while***

7: ***return*** $a(t)$, *regret*$_t$ ▷ *get a(t) and the regret*

8: ***end procedure***

Proposition 7. *If* $\hat{b} \leq \min_l c_l(\bar{a}_l - \underline{a}_l)$ *then Algorithm* (18) *reduces to*

$$\begin{cases} a_l := \bar{a}_l x_l + \underline{a}_l (1 - x_l), \\ \dot{x}_l = x_l \left[\left\langle e_l, \hat{f}(a) \right\rangle - 1\hat{b} \sum_l \left\langle e_l, \hat{f}(a) \right\rangle x_l [c_l(\bar{a}_l - \underline{a}_l)] \right], \\ l \in \{1, \ldots, n\} \end{cases} \tag{19}$$

Proof. It suffices to check that for $\hat{b} \leq \min_l c_l(\bar{a}_l - \underline{a}_l)$, the vector z defined by $z_l = \frac{e^{y_l}}{\sum_{k=1}^{n} e^{y_k}}$ solves the replicator equation,

$$\dot{z}_l = z_l [\dot{y}_l - \langle z, \dot{y} \rangle].$$

Thus, $x_l = \frac{e^{y_l}}{\sum_{k=1}^{n} e^{y_k}} \frac{\hat{b}}{[c_l(\bar{a}_l - \underline{a}_l)]}$ solves $\dot{x}_l = x_l \left[\left\langle e_l, \hat{f}(a) \right\rangle - 1\hat{b} \sum_l \left\langle e_l, \hat{f}(a) \right\rangle x_l [c_l(\bar{a}_l - \underline{a}_l)] \right]$. This completes the proof.

Note that the dynamics of x in Eq. (19) is a constrained replicator dynamics [8] which is widely used in evolutionary game dynamics. This observation establishes a relationship between optimization and game dynamics and explains that the replicator dynamics is the gradient flow of the (expected payoff) under simplex constraint.

The next example illustrates a constrained distributionally robust optimization in wireless communication networks.

Example 5 (Wireless communication). *Consider a power allocation problem over n medium access channels. The signal-to-interference-plus-noise ratio (SINR) is*

$$SINR_l = \frac{\frac{a_l|\omega_{ll}|^2}{\left(d^2(s_r(l),s_t(l))+\varepsilon^2\right)^{\frac{o}{2}}}}{N_0(s_r(l)) + I_l(s_r(l))},$$

where

- $N_0 > 0$ is the background noise.
- The interference on channel l is denoted $I_l \geq 0$. One typical model for I_l is

$I_l = \sum_{k\neq l} \frac{a_k|\omega_{kl}|^2}{\left(d^2(s_r(l),s_t(k))+\varepsilon^2\right)^{\frac{o}{2}}}.$

- $\epsilon > 0$ is the height of the transmitter antenna.
- ω_{ll} is the channel state at l. The channel state is unknown. Its true distribution is also unknown.
- $s_r(l)$ is the location of the receiver of l
- $s_t(l)$ is the location of the transmitter of l
- $o \in \{2, 3, 4\}$ is the pathloss exponent.
- a_l is the power allocated to channel l. It is assumed to be between $\underline{a}_l \geq 0$ and $\bar{a}_l$ with $0 \leq \underline{a}_l < \bar{a}_l < +\infty$. Moreover, a total power budget constraint is imposed $\sum_{l=1}^{n} a_l \leq \bar{a}$ where $\bar{a} > \sum_{l=1}^{n} \underline{a}_l \geq 0$.

It is worth mentioning that the action constraint of the power allocation problem are similar to the ones analyzed in Section 3. The admissible action space is

$$\mathcal{A} := \left\{ a \in \mathbb{R}_+^n : \ \underline{a}_l \leq a_l \leq \bar{a}_l, \sum_{l=1}^{n} a_l \leq \bar{a} \right\}.$$

Clearly, $\mathcal{A}$ is a non-empty convex compact set. The payoff function is the sum-rate $r(a,\omega) = \sum_{l=1}^{n} W_l \log(1 + SINR_l)$ where $W_l > 0$. The mapping $(a,\omega) \mapsto r(a,\omega)$ is continuously differentiable.

- Robust *optimization is too conservative: Part of the robust optimization problem* [9, 7] *consists of choosing the channel gain $|\omega_{ll}|^2 \in [0, \bar{\omega}_{ll}]$ were the bound $\bar{\omega}$ need to be carefully designed. However the worst case is achieved when the channel gain is zero: $\inf_{\omega \in \prod_l [0,\bar{\omega}_{ll}]} r(a,\omega) = 0$. Hence the robust performance is zero. This is too conservative as several realizations of the channel may give better performance than zero. Another way is to re-design the bounds $\underline{\omega}_{ll}$ and $\bar{\omega}_{ll}$. But if $\underline{\omega}_{ll} > 0$ it means that very low channel gains are not allowed, which may be too optimistic. Below we use the distributional robust optimization approach which eliminates this design issue.*
- Distributional *robust optimization: By means of the training sequence or channel estimation method, a certain (statistical) distribution m is derived. However m cannot be considered as the*

true distribution of the channel state due to estimation error. The true distribution of ω is unknown. Based on this observation, an uncertainty set $B_\rho(m)$ with radius $\rho \geq 0$ is constructed for alternative distribution candidates. Note that $\rho = 0$ means that $B_0(m) = \{m\}$. The distributional robust optimization problem is $\sup_a \inf_{\tilde{m} \in B_\rho(m)} \mathbb{E}_{\tilde{m}} r(a, \omega)$. In presence of interference, the function $r(a, \omega)$ is not necessarily concave in a. In absence of interference, the problem becomes concave.

4. Distributed optimization

This section presents distributed distributionally robust optimization problems over a direct graph. A large number of virtual agents can potentially choose a node (vertex) subject to constraint. The vector a represents the population state. Since a has n components, the graph has n vertices. The interactions between virtual agents are interpreted as possible connections of the graph. Let us suppose that the current interactions are represented by a directed graph $\mathcal{G} = (\mathcal{L}, \mathcal{E})$, where $\mathcal{E} \subseteq \mathcal{L}^2$ is the set of links representing the possible interaction among the proportion of agents, i.e., if $(l, k) \in \mathcal{E}$, then the component l of a can interact with the $k-$th component of a. In other words, $(l, k) \in \mathcal{E}$ means that virtual agents selecting the strategy $l \in \mathcal{L}$ could migrate to strategy $k \in \mathcal{L}$. Moreover, $\Lambda \in \{0, 1\}^{n \times n}$ is the adjacency matrix of the graph $\mathcal{G}$, and whose entries are $\lambda_{lk} = 1$, if $(l, k) \in \mathcal{E}$; and $\lambda_{lk} = 0$, otherwise.

Definition 6. *The distributionally robust fitness function is the marginal distributionally robust payoff function. If $a \mapsto \mathbb{E}_m h(a, \omega)$ is continuously differentiable, the distributionally robust fitness function is $\mathbb{E}_m \nabla_a h(a, \omega)$.*

Definition 7. *The virtual population state a is an equilibrium if $a \in \mathcal{A}$ and it solves the variational inequality*

$$\langle a - b, \mathbb{E}_m \nabla_a h(a, \omega) \rangle \geq 0, \quad \forall b \in \mathcal{A}.$$

Proposition 8. *Let the set of virtual population state $\mathcal{A}$ be non-empty convex compact and $b \mapsto \mathbb{E}_m \nabla h(b, \omega)$ be continuous. Then the following conditions are equivalent:*

- $\langle a - b, \mathbb{E}_m \nabla h(a, \omega) \rangle \geq 0, \quad \forall b \in \mathcal{A}.$
- *the action a satisfies* $a = proj_{\mathcal{A}}[a + \eta \mathbb{E}_m \nabla h(a, \omega)]$

Proof. Let a be a feasible action that solves the variational inequality:

$$\langle a - b, \mathbb{E}_m \nabla h(a, \omega) \rangle \geq 0, \quad \forall b \in \mathcal{A}.$$

Let $\eta > 0$. By multiplying both sides by η, we obtain

$$\langle a - b, \eta \mathbb{E}_m \nabla h(a, \omega) \rangle \geq 0, \quad \forall b \in \mathcal{A}.$$

We add the term $\langle a, b-a\rangle$ to both sides to obtain the following relationships:

$$\begin{aligned}
&\langle a-b, \eta\mathbb{E}_m\nabla h(a,\omega)\rangle \geq 0 && \forall b\in\mathcal{A},\\
&\Leftrightarrow \langle a-b, \eta\mathbb{E}_m\nabla h(a,\omega)\rangle + \langle a-b, -a\rangle \geq \langle a, b-a\rangle && \forall b\in\mathcal{A},\\
&\Leftrightarrow \langle b-a, -[a+\eta\mathbb{E}_m\nabla h(a,\omega)]\rangle + \langle a-b, -a\rangle \geq 0 && \forall b\in\mathcal{A},\\
&\Leftrightarrow \langle b-a, a-[a+\eta\mathbb{E}_m\nabla h(a,\omega)]\rangle \geq 0 && \forall b\in\mathcal{A},
\end{aligned} \tag{20}$$

Recall that the projection operator on a convex and closed set $\mathcal{A}$ is uniquely determined by

$$z\in\mathbb{R}^n, z' = proj_{\mathcal{A}}[z] \Leftrightarrow \langle z'-z, b-z'\rangle \geq 0, \ \ \forall b\in\mathcal{A}.$$

Thus

$$\begin{aligned}
&\langle b-a, a-[a+\eta\mathbb{E}_m\nabla h(a,\omega)]\rangle \geq 0, \ \ \forall b\in\mathcal{A}\\
&\Leftrightarrow a = proj_{\mathcal{A}}[a+\eta\mathbb{E}_m\nabla h(a,\omega)].
\end{aligned} \tag{21}$$

This completes the proof.

As a consequence we can derive the following existence result.

Proposition 9. *Let the set of virtual population states $\mathcal{A}$ be a non-empty convex compact and the mapping $b\mapsto\mathbb{E}_m\nabla h(b,\omega)$ be continuous. Then, there exists at least one equilibrium in $\mathcal{A}$.*

Proof. A direct application of the Brouwer-Schauder's fixed-point theorem which states that if $\phi: \ \mathcal{A}\to\mathcal{A}$ is continuous and $\mathcal{A}$ non-empty convex compact then ϕ has at least one fixed-point in $\mathcal{A}$. Here we choose $\phi(a) = proj_{\mathcal{A}}[a+\eta\mathbb{E}_m\nabla h(a,\omega)]$. Clearly $\phi(\mathcal{A})\subseteq\mathcal{A}$ and ϕ is continuous on $\mathcal{A}$ as the mapping $b\mapsto\mathbb{E}_m\nabla h(b,\omega)$ and the projection operator $b\mapsto proj_{\mathcal{A}}[b]$ are both continuous. Then the announced result follows. This completes the proof.

Note that we do not need sophisticated set-valued fixed-point theory to obtain this result.

Definition 8. *The virtual population state a is evolutionarily stable if $a\in\mathcal{A}$ and for any alternative deviant state $b\neq a$ there is an invasion barrier $\epsilon_b > 0$ such that*

$$\langle a-b, \mathbb{E}_m\nabla h(a+\epsilon(b-a),\omega) > 0, \ \ \forall\epsilon\in(0,\epsilon_b).$$

The function $\varrho : \mathcal{A}\times\mathbb{R}^n\times\mathbb{R}_+^{n\times n}\to\mathbb{R}^{n\times n}$ is the revision protocol, which describes how virtual agents are making decisions. The revision protocol ϱ takes a population state a, the corresponding fitness $\nabla\mathbb{E}_m h$, the adjacency matrix Λ and returns a matrix. Therefore, let $\varrho_{lk}(a,h,\Lambda)$ be the switching rate from the lth to k^{th} component. Then, the virtual agents selecting the strategy $l\in\mathcal{L}$ have incentives to migrate to the strategy $l\in\mathcal{L}$ only if $\varrho_{lk}(a,h,\Lambda) > 0$, and it is also possible to design switch rates depending on the topology describing the migration constraints, i.e., $\lambda_{lk} = 0 \Rightarrow \varrho_{lk}(a,h,\Lambda) = 0$. The distributed distributionally robust optimization consists to perform the optimization problem above over the distributed network that is subject to communication restriction. We construct a distributed distributionally robust game dynamics to perform such a task. The distributed distributionally robust evolutionary game dynamics emerge from the combination of the (robust) fitness h and the constrained switching rates ϱ. The evolution of the portion a_l is given by the distributed distributional robust mean dynamics

$$\dot{a}_l = \sum_{k \in \mathcal{L}} a_k \varrho_{kl}(a, h, \Lambda) - a_l \sum_{k \in \mathcal{L}} \varrho_{lk}(a, h, \Lambda), l \in \mathcal{L}, \tag{22}$$

Since the distributionally robust function h is obtained after the transformation from payoff function r by means of triality theory, the dynamics (22) is seeking for distributed distributionally robust solution.

Algorithm 5. *The distributed distributional robust mean dynamics pseudocode is as follows:*

1: ***procedure*** POPULATION-INSPIRED ALGORITHM $(a(0), \epsilon, T, \varrho, g, m, h, \Lambda)$ ⊳ *The population-inspired learning starting from $a(0)$ within $[0, T]$*

2: $a \leftarrow a(0)$

3: ***while*** *regret* $> \epsilon$ *and* $t \leq T$ ***do*** ⊳ *We have the answer if regret is 0*

4: *Compute $a(t)$ solution of (22)*

5: *Compute $regret_t$*

6: ***end while***

7: ***return*** $a(t)$, $regret_t$ ⊳ *get $a(t)$ and the regret*

8: ***end procedure***

The next example establishes evolutionarily stable state, equilibria and rest-point of the dynamics (22) by designing ϱ.

Example 6. *Let us consider a power system that is composed of* 10 *generators, i.e., let $\mathcal{L} = \{1, \ldots, 10\}$. Let $a_l \in \mathbb{R}_+$ be the power generated by the generator $l \in \mathcal{L}$. Each power generation should satisfy the physical and/or operation constraints $a_l \in [\underline{a}_l, \bar{a}_l]$, for all $l \in \mathcal{L}$. It is desired to satisfy the power demand given by $d \in \mathbb{R}$, i.e., it is necessary to guarantee that $\sum_{l \in \mathcal{L}} a_l = d$, i.e., the supply meets the demand. The objective is to minimize the generation quadratic costs for all the generators, i.e.,*

$$\text{Maximize } r(a, \omega) = \sum_{l \in \mathcal{L}} r_l(a_l) = -\sum_{l \in \mathcal{L}} \left(c_{0l} + c_{1l} a_l + c_{2l} a_l^2\right),$$

$$\text{s.t.} \quad \sum_{l \in \mathcal{L}} a_l = d, \quad \underline{a}_l \leq a_l \leq \bar{a}_l, \ l \in \mathcal{L},$$

where $r : \mathbb{R}^n \to \mathbb{R}$ is concave, and the parameters are possibly uncertain and selected as $c_{0l} = 25 + 6l$, $c_{1l} = 15 + 4l + \omega_{1l}$, $c_{2l} = 5 + l + \omega_{2l}$, and $d = 20 + \omega_{3l}$. Therefore, the fitness functions for the corresponding full potential game are given by $f_l(a) = -2a_l c_{2l} - c_{1l}$, for all $l \in \mathcal{L}$, and action space is given by

$$\mathcal{A} = \left\{ a \in \mathbb{R}^n_+ : \sum_{l \in \mathcal{L}} a_l = d, \;\; a_l \in \left[\underline{a}_l, \bar{a}_l\right] \right\}.$$

The distributed revision protocol is set to

$$\varrho_{lk}(a, h, \Lambda) = \frac{\lambda_{lk}}{a_l} \max(0, \bar{a}_k - a_k) \max\left(0, a_l - \underline{a}_l\right) \max(0, \mathbb{E}_m(h_k - h_l)),$$

for $a_l \neq 0$. We evaluate four different scenarios, i.e.,

1. $\underline{a} = 0_n$ *and* $\bar{a} = d\mathbb{1}_n$,
2. $\underline{a}_l = 0$, *for all* $l \in \mathcal{L}\ \{9, 10\}$, $\underline{a}_9 = 1.1$, *and* $\underline{a}_{10} = 1$; *and* $\bar{a}_l = d$, *for all* $l \in \mathcal{L}\ \{1, 2\}$, $\bar{a}_1 = 3$, *and* $\bar{a}_2 = 2.5$,
3. *Case 1 constraints and with interaction restricted to the cycle graph* $\mathcal{G} = (\mathcal{L}, \mathcal{E})$ *with set of links* $\mathcal{E} = \left\{ \cup_{l \in \mathcal{L}\ \{n\}} (l, l+1) \right\} \cup \{(n, 1)\}$,
4. *Case 2 constraints and with interaction restricted as in Case 3.*

Figure 3 presents the evolution of the generated power, the fitness functions corresponding to the marginal costs and the total cost. For the first scenario, the evolutionary game dynamics converge to a standard evolutionarily stable state in which $\hat{f}(a^\star) = c\mathbb{1}_n$. In contrast, for the second scenario, the dynamics converge to a constrained evolutionarily stable state.

4.1. Extension to multiple decision-makers

Consider a constrained game $\mathcal{G}$ in strategic-form given by

- $\mathcal{P} = \{1, \ldots, P\}$ is the set of players. The cardinality of $\mathcal{P}$ is $P \geq 2$.
- Player p has a decision space $A_p \subset \mathbb{R}^{n_p}$, $n_p \geq 1$. Players are coupled through their actions and their payoffs. The set of all feasible action profiles is $\mathcal{A} \subset \mathbb{R}^n$, with $n = \sum_{p \in \mathcal{P}} n_p$. Player p can choose an action a_p in the set $\mathcal{A}_p(a_{-p}) = \left\{ a_p \in A_p : \;\; \left(a_p, a_{-p}\right) \in \mathcal{A} \right\}$.
- Player p has a payoff function $r_p : \;\; \mathcal{A} \to \mathbb{R}$.

We restrict our attention to the following constraints:

$$A_p = \left\{ a_p \in \mathbb{R}^{n_p} \mid a_{pl} \in \left[\underline{a}_{pl}, \bar{a}_{pl}\right], \;\; l \in \{1, \ldots, n_p\}, \;\; \sum_{l=1}^{n_p} c_{pl} a_{pl} \leq b_p \right\}$$

The coupled constraint is

$$\mathcal{A} = \left\{ a \in \prod_p A_p, \;\;\; \sum_{p \in \mathcal{P}} \left\langle \bar{c}_p, a_p \right\rangle \leq \bar{b} \right\}.$$

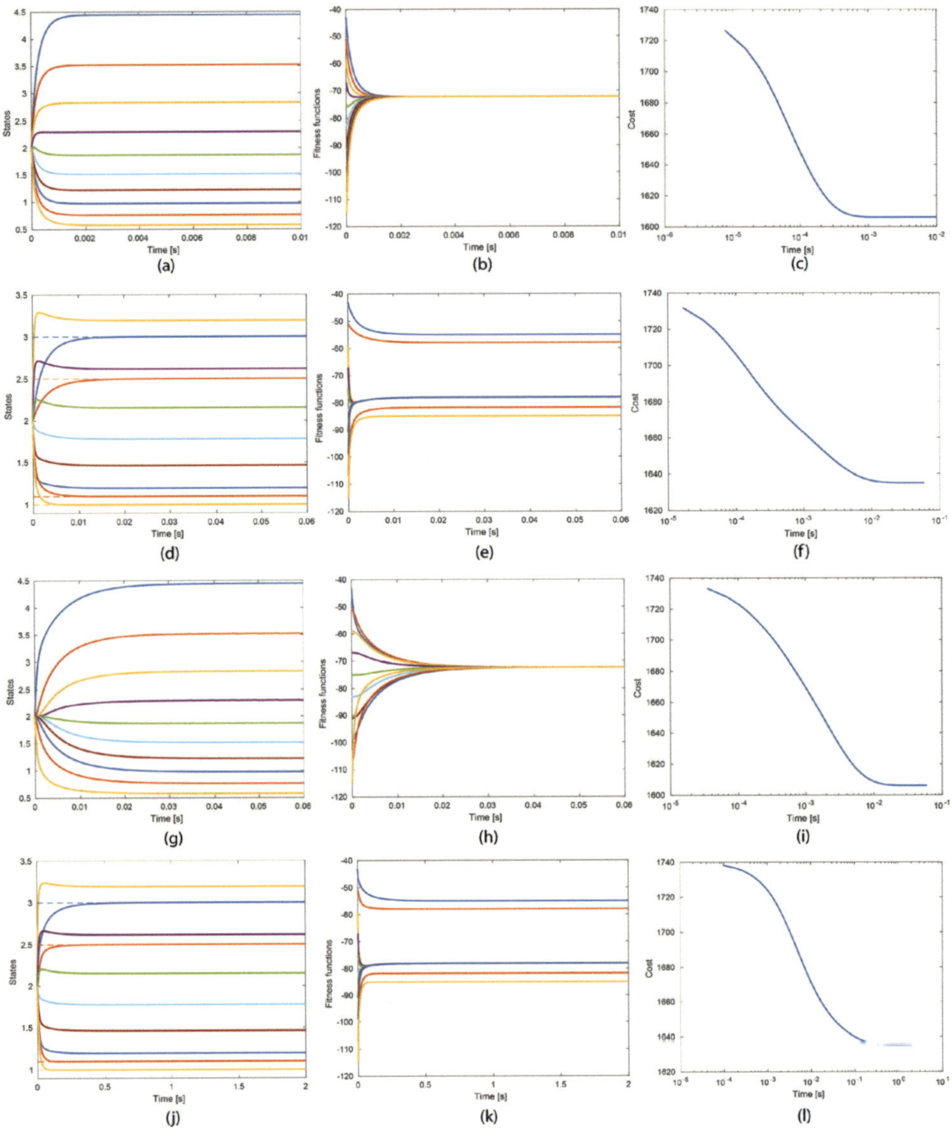

Figure 3. Economic power dispatch. Evolution of the population states (generated power), fitness functions $\hat{f}(a) = \nabla \mathbb{E}h(a,\omega)$, and the costs $-\mathbb{E}r(a,\omega)$. Figures (a)-(c) for case 1, (d)-(f) for case 2, (g)-(i) for case 3, and (j)-(l) for case 4.

Feasibility condition: If $\underline{a}_{pl} < \bar{a}_{pl}, \ \ l \in \{1,\dots,n_p\}, c_{pl} > 0, \ \ \sum_{l=1}^{n_p} c_{pl}\underline{a}_{pl} < b_p, \ \ \bar{c}_p \in \mathbb{R}^{n_p}_{>0}$ and $\sum_{p\in\mathcal{P}} \left\langle \bar{c}_p, \underline{a}_p \right\rangle < \bar{b},$ the constraint set $\mathcal{A}$ is non-empty, convex and compact.

We propose a method to compute a constrained equilibrium that has a full support (whenever it exists). We do not use the projection operator. Indeed we transform the domain $\left[\underline{a}_{pl}, \bar{a}_{pl}\right] = \xi([0,1])$ where $\xi(x_{pl}) = \bar{a}_{pl}x_{pl} + \underline{a}_{pl}(1 - x_{pl}) = a_{pl}$. ξ is a one-to-one mapping and

$$x_{pl} = \xi^{-1}(a_{pl}) = \frac{a_{pl} - \underline{a}_{pl}}{\overline{a}_{pl} - \underline{a}_{pl}} \in [0, 1].$$

$$\sum_{l=1}^{n_p} c_{pl}\left(\overline{a}_{pl} - \underline{a}_{pl}\right)x_{pl} \leq b_p - \sum_{l=1}^{n_p} c_{pl}\underline{a}_{pl} =: \hat{b}_p.$$

The learning algorithm (23) is

$$\begin{cases} \dot{y}_p = \left[\nabla_p^2 g\right]^{-1} \nabla_{a_p} r_p(a, \omega), \\ a_{pl} := \overline{a}_{pl} x_{pl} + \underline{a}_{pl}(1 - x_{pl}), \\ x_{pl} = \min\left(1, \frac{e^{y_{pl}}}{\sum_{k=1}^{n_p} e^{y_{pk}}} \frac{\hat{b}_p}{\left[c_{pl}\left(\overline{a}_{pl} - \underline{a}_{pl}\right)\right]}\right), \\ l \in \{1, \ldots, n_p\}, \end{cases} \tag{23}$$

generates a trajectory $a_p(t) = \left(a_{pl}(t)\right)_l$ that satisfies the constraint of player p at any time t.

5. Notes

The work in [10] provides a nice intuitive introduction to robust optimization emphasizing the parallel with static optimization. Another nice treatment [11], focusing on robust empirical risk minimization problem, is designed to give calibrated confidence intervals on performance and provide optimal tradeoffs between bias and variance [12, 13]. f-divergence based performance evaluations are conducted in [11, 14, 15]. The connection between risk-sensitivity measures such as the exponentiated payoff and distributionally robustness can be found in [16]. Distributionally robust optimization and learning are extended to multiple strategic decision-making problems i.e., distributionally robust games in [17, 18].

Acknowledgements

We gratefully acknowledge support from U.S. Air Force Office of Scientific Research under grant number FA9550-17-1-0259.

Author details

Jian Gao, Yida Xu, Julian Barreiro-Gomez, Massa Ndong, Michalis Smyrnakis and Hamidou Tembine*

*Address all correspondence to: tembine@ieee.org

Learning and Game Theory Laboratory, New York University, Abu Dhabi, United Arab Emirates

References

[1] Sion M. On general minimax theorems. Pacific Journal of Mathematics. 1958;**8**(1):171-176

[2] Bach FR. Duality between subgradient and conditional gradient methods. SIAM Journal on Optimization. 2015;**1**(25):115-129

[3] Kim D, Fessler JA. Optimized first-order methods for smooth convex minimization. Mathematical Programming. 2016;**159**(1):81-107

[4] Nesterov Y. Accelerating the cubic regularization of newton's method on convex problems. Mathematical Programming. 2008;**112**(1):159-181

[5] Nesterov Y. Primal-dual subgradient methods for convex problems. Mathematical Programming. 2009;**120**(1):221-259

[6] Larry A. Minimization of functions having Lipschitz continuous first partial derivatives. Pacific Journal of Mathematics. Tokyo Japan: International Academic Printing Co., Ltd.; 1966;**16**(1):1-3

[7] Tembine H. Distributed Strategic Learning for Wireless Engineers. Boca Raton, FL, USA: CRC Press, Inc.; 2012. p. 496

[8] Taylor PD, Jonker L. Evolutionarily stable strategies and game dynamics. Mathematical Biosciences. 1978;**40**:145-156

[9] Ben-Tal A, Ghaoui LE, Nemirovski A. Robust Optimization. Princeton Series in Applied Mathematics. Princeton University Press; August 30, 2009. 576 p. ISBN-10: 0691143684, ISBN-13: 978-0691143682

[10] Esfahani PM, Kuhn D. Data-driven distributionally robust optimization using the Wasserstein metric: Performance guarantees and tractable reformulations. In: Mathematical Programming. Jul 2017

[11] Duchi JC, Namkoong H. Stochastic gradient methods for distributionally robust optimization with f-divergences. Advances in Neural Information Processing Systems. 2015; 2208-2216

[12] Ben-Tal A, den Hertog D, Waegenaere AD, Melenberg B, Rennen G. Robust solutions of optimization problems affected by uncertain probabilities. Management Science. Catonsville, USA: Informs PubsOnLine; 2013;**59**(2):341-357

[13] Ben-Tal A, Hazan E, Koren T, Mannor S. Oracle-based robust optimization via online learning. Operations Research. 2015;**63**(3):628-638

[14] Ahmadi-Javid A. An information-theoretic approach to constructing coherent risk measures. IEEE International Symposium on Information Theory Proceedings. 2011;2125-2127

[15] Ahmadi-Javid A. Entropic value-at-risk: A new coherent risk measure. Journal of Optimization Theory and Applications. 2012;**155**(3):1105-1123

[16] Föllmer H, Knispel T. Entropic risk measures: Coherence vs. convexity, model ambiguity, and robust large deviations. Stochastics Dynamics. 2011;**11**:333-351

[17] Bauso D, Gao J, Tembine H. Distributionally robust games: F-divergence and learning. ValueTools, International Conference on Performance Evaluation Methodologies and Tools, Venice, Italy; December 5-7, 2017

[18] Tembine H. Dynamic robust games in mimo systems. IEEE Transactions on Systems, Man, and Cybernetics, Part B (Cybernetics). Aug 2011;**41**(4):990-1002

Polyhedral Complementarity Approach to Equilibrium Problem in Linear Exchange Models

Vadim I. Shmyrev

Additional information is available at the end of the chapter

http://dx.doi.org/10.5772/intechopen.77206

Abstract

New development of original approach to the equilibrium problem in a linear exchange model and its variations is presented. The conceptual base of this approach is the scheme of polyhedral complementarity. The idea is fundamentally different from the well-known reduction to a linear complementarity problem. It may be treated as a realization of the main idea of the linear and quadratic programming methods. In this way, the finite algorithms for finding the equilibrium prices are obtained. The whole process is a successive consideration of different structures of possible solution. They are analogous to basic sets in the simplex method. The approach reveals a decreasing property of the associated mapping whose fixed point yields the equilibrium of the model. The basic methods were generalized for some variations of the linear exchange model.

Keywords: exchange model, economic equilibrium, fixed point, polyhedral complementarity, optimization problem, conjugate function, algorithm

1. Introduction

It is known that the problem of finding an equilibrium in a linear exchange model can be reduced to the linear complementarity problem [1]. Proposed by the author in [2], a polyhedral complementarity approach is based on a fundamentally different idea that reflects more the character of economic equilibrium as a concordance the consumers' preferences with financial balances. In algorithmic aspect, it may be treated as a realization of the main idea of linear and quadratic programming. It has no analogues and makes it possible to obtain the finite algorithms not only for the general case of classical linear exchange model [3], but also for more complicate linear models, in which there are two sets of participants: consumers and firms producing goods [4] (more references one can find in [5]). The simplest algorithms are those for a model with fixed budgets, known more as Fisher's problem. The convex programming reduction of it, given by Eisenberg and Gale [6], is well known. This result has been used by many authors to study computational aspects of the problem. Some reviews of that can be found in [7]. The polyhedral complementarity approach gives an alternative reduction of the Fisher's problem to a convex program [2, 8]. Only the well-known elements of transportation problem algorithms are used in the procedures obtained by this way [9]. These simple procedures can be used for getting iterative methods for more complicate models [5, 10].

The mathematical fundamental base of the approach is a special class of piecewise constant multivalued mappings on the simplex in $\mathbb{R}^n$, which possesses some monotonicity property (decreasing mappings). The problem is to find a fixed point of the mapping. The mappings in the Fisher's model proved to be potential ones. This makes it possible to reduce a fixed point problem to two optimization problems which are in duality similarly to dual linear programming problems. The obtained algorithms are based on the ideas of suboptimization [11]. The mapping for the general exchange model is not potential. The proposed finite algorithm can be considered as an analogue of the Lemke's method for linear complementarity problem with positive principal minors of the restriction matrix (class P) [12].

2. Polyhedral complementarity problem

The basic scheme of the considered approach is the polyhedral complementarity. We consider polyhedrons in $\mathbb{R}^n$. Let two polyhedral complexes ω and ξ with the same number of cells r be given. Let $R \subset \omega \times \xi$ be a one-to-one correspondence: $R = \{(\Omega_i, \Xi_i)\}_{i=1}^r$ with $\Omega_i \in \omega$, $\Xi_i \in \xi$.

We say that the complexes ω and ξ are *in duality by R* if the subordination of cells in ω and the subordination of the corresponding cells in ξ are opposite to each other:

$$\Omega_i \prec \Omega_j \iff \Xi_i \succ \Xi_j.$$

The polyhedral complementarity problem is to find a point that belongs to both cells of some pair (Ω_l, Ξ_l):

$$p^* \text{ is the solution } \iff p^* \in \Omega_i \cap \Xi_i \text{ for some } i.$$

This is natural generalization of linear complementarity, where (in nonsingular case) the complexes are formed by all faces of two simplex cones.

Figure 1 shows an example of the polyhedral complementarity problem. Each of two complexes has seven cells. There is a unique solution of the problem—the point x^* that belongs to Ω_6 and Ξ_6.

The polyhedral complementarity problem can be reformulated as a fixed point one. To do this the *associated mapping* is introduced as follows:

$$G(p) = \Xi_i \quad \forall p \in \Omega_i^\circ,$$

where Ω_i° is the relative interior of Ω_i.

Now p^* is the solution of complementarity problem if $p^* \in G(p^*)$.

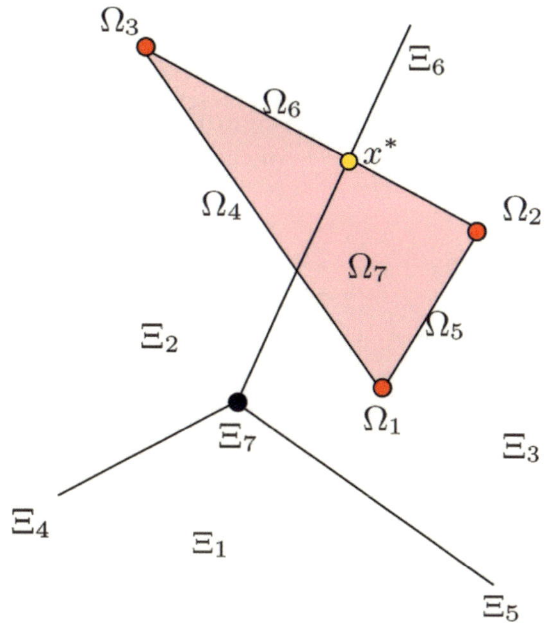

Figure 1. Polyhedral complementarity.

3. Classical linear exchange model

We demonstrate the main idea of the approach on the classical linear exchange model in the well-known description [13].

Consider a model with n commodities (goods) and m consumers. Let $J = \{1, \ldots, n\}$ and $I = \{1, \ldots, m\}$ be the index sets of commodities and consumers.

Each consumer $i \in I$ possesses a vector of initial endowments $w^i \in R^n_+$. The exchange of commodities is realized with respect to some nonnegative prices p_j, forming a price vector $p \in R^n_+$.

The consumer $i \in I$ has to choose a consumption vector $x^i \in R^n_+$ maximizing his linear utility function (c^i, x^i) under *budget restriction*:

$$\left.\begin{array}{c}(c^i, x^i) \to \max,\\ (p, x^i) \leqslant (p, w^i),\\ x^i \geqslant 0.\end{array}\right| \Longrightarrow \text{The problem of consumer } i.$$

Let $\tilde{x}^i$ be a vector x^i that solves this program.

A price vector $\tilde{p} \neq 0$ is an *equilibrium price vector* if there exist solutions $\tilde{x}^i$, $i = 1, \ldots, m$, for the individual optimization problems such that

$$\sum_{i \in I} \tilde{x}^i = \sum_{i \in I} w^i.$$

In what follows, we normalize the initial endowment of each commodity to 1, that is, $\sum_i w^i = (1, \ldots, 1) \in R^n$. The sum of p_j is also normalized to 1, restricting the price vector p to lie in the unit simplex

$$\sigma = \left\{ p \in R^n_+ \mid \sum_{j \in J} p_j = 1 \right\}.$$

For the sake of simplicity assume $c^i > 0$, $\forall i \in I$. It is sufficient for existence of equilibrium [13].

4. The main idea of the approach

The equilibrium problem can be considered in two different ways.

1°. The traditional point of view: supply–demand balance.

Given a price vector p, the economy reacts by supply and demand vectors:

$$p \begin{array}{l} \nearrow \text{ demand } D(p) \\ \searrow \text{ supply } S(p) \end{array}.$$

The goods' balance is the condition of equilibrium:

$$\hat{p} \text{ is equilibrium price vector } \Longleftrightarrow S(\hat{p}) = D(\hat{p}).$$

2°. Another point of view.

The presented consideration is based on the new notion of consumption *structure*.

Definition. *A set $B \subset I \times J$ is named a structure, if for each $i \in I$ there exists $(i, j) \in B$.*

Say that a consumption prescribed by $\{x^i\}$ is consistent with structure B if

$$(i, j) \notin B \quad \Longrightarrow \quad x^i_j = 0.$$

This notion is analogous to the basic index set in linear programming.

Two sets of the price vectors can be considered for each structure B.

We name them *zones*:

$$B \begin{array}{l} \nearrow \text{ the } \textit{preference zone } \Xi(B) \\ \searrow \text{ the } \textit{balance zone } \Omega(B). \end{array}$$

$\Xi(B)$ is the set of prices by which the consumers prefer the connections of the structure, ignoring the budget conditions and balances of goods. $\Omega(B)$ is the set of prices by which the budget conditions and balances of goods are possible when the connections of the structure are respected, but the participants' preferences are ignored.

Now, it is clear that

$$p \text{ is an equlibrium price vector } \Longleftrightarrow (\exists \mathcal{B}) p \in \Omega(\mathcal{B}) \cap \Xi(\mathcal{B}).$$

We show that in this way the equilibrium problem is reduced to polyhedral complementarity one.

The question is as follows: *What kind of the structures $\mathcal{B} \in \mathfrak{B}$ should be considered and what should be the collection $\mathfrak{B}$?*

3°. The parametric transportation problem of the model.

Given a price vector p consider the following *transportation problem of the model:*

$$\sum_{i \in I} \sum_{j \in J} z_{ij} \ln c_j^i \to \max$$

under conditions

$$\{z_{ij}\} \in Z(p) \left| \begin{array}{ll} \sum_{j \in J} z_{ij} = (p, w^i), & i \in I, \\ \sum_{i \in I} z_{ij} = p_j, & j \in J, \\ z_{ij} \geqslant 0, & (i,j) \in I \times J. \end{array} \right.$$

The equations of this problem represent the financial balances for the consumers and commodities. The variables z_{ij} are introduced by $z_{ij} = p_j x_j^i$.

This is the classical transportation problem. The price vector p is a parameter of the problem. Under the assumption about $\{w^i\}$ this problem is solvable for each $p \in \sigma$.

The answer on the question about $\mathfrak{B}$ reads: *$\mathfrak{B}$ is the collection of all dual feasible basic index sets of the transportation problem and of all their subsets being structures.*

4°. Polyhedral complexes of the model.

For $\mathcal{B} \in \mathfrak{B}$, we obtain the description of zones $\Omega(\mathcal{B})$ and $\Xi(\mathcal{B})$ in the following way.

$$\mathcal{B} \in \mathfrak{B} \Rightarrow \left| \begin{array}{l} a) \quad \Omega(\mathcal{B}) \subset \sigma \text{ is the balance zone of the structure :} \\ \Omega(\mathcal{B}) = \{p \in \sigma | \exists z \in Z(p), z_{ij} = 0, (i,j) \notin \mathcal{B}\}; \\ b) \quad \Xi(\mathcal{B}) \subset \sigma^\circ \text{ is the preferance zone of the structure :} \\ \qquad \Xi(\mathcal{B}) = \left\{ q \in \sigma^\circ \left| \max_k \frac{c_k^i}{q_k} = \frac{c_j^i}{q_j}, \ \forall (i,j) \in \mathcal{B} \right. \right\}. \end{array} \right.$$

Here, σ° is the relative interior of σ.

It is easy to give these descriptions in more detail.

For $q \in \Xi(\mathcal{B})$, we have the linear system

$$\frac{q_k}{c_k^i} = \frac{q_j}{c_j^i} \qquad (i,k) \in \mathcal{B}, (i,j) \in B, \tag{1}$$

$$\frac{q_l}{c_l^i} \geqslant \frac{q_j}{c_j^i} \qquad (i,l) \notin \mathcal{B}, (i,j) \in B. \tag{2}$$

Thus, $\Xi(\mathcal{B})$ is the intersection of a polyhedron with σ°.

To obtain the description of $\Omega(\mathcal{B})$, we should use the well-known tools of transportation problems theory. Given $B \in \mathfrak{B}$, introduce a graph $\Gamma(\mathcal{B})$ with the set of vertices $V = \{1, 2, \ldots, m+n\}$ and the set of edges $\{(i, m+j) | (i,j) \in \mathcal{B}\}$. Let τ be the number of components of this graph, let V_ν be the set of vertices of ν th component, $I_\nu = I \cap V_\nu$ and $J_\nu = \{j \in J | (m+j) \in V_\nu\}$.. It is not difficult to show that the following system of linear equations must hold for $p \in \Omega(\mathcal{B})$:

$$\sum_{j \in J_\nu} p_j = \sum_{i \in I_\nu} (p, w^i), \qquad \nu = 1, \ldots, \tau. \tag{3}$$

Under these conditions, the values z_{ij} can be obtained from the conditions $z \in Z(p)$ and

$$z_{ij} = 0, \qquad (i,j) \notin B,$$

presenting linear functions of p: $z_{ij} = z_{ij}(p)$. Now, for $p \in \Omega(\mathcal{B})$, we have in addition the system of linear inequalities

$$z_{ij}(p) \geqslant 0, \qquad (i,j) \in \mathcal{B}.$$

Thus, $\Omega(\mathcal{B})$ is described by a linear system of equalities and inequalities. Therefore, it is also a polyhedron.

It is easy to see that each face of the polyhedron $\Omega(\mathcal{B})$ is also a polyhedron $\Omega\left(\mathcal{B}'\right)$ with $\mathcal{B}' \subset \mathcal{B}$. Therefore, we have on the simplex σ a *polyhedral complex* $\omega = \{\Omega(\mathcal{B}) | \mathcal{B} \in \mathfrak{B}\}$. The polyhedrons $\Xi(\mathcal{B})$ form on σ another polyhedral complex $\xi = \{\Xi(\mathcal{B}) | \mathcal{B} \in \mathfrak{B}\}$. It is clear that

$$\Omega(\mathcal{B}_1) \subset \Omega(\mathcal{B}_2) \quad \Longrightarrow \quad \Xi(\mathcal{B}_1) \supset \Xi(\mathcal{B}_2).$$

Thus, the complexes ω, ξ are in duality, and we obtain the reduction of the equilibrium problem to a polyhedral complementarity one.

Example. In the model, there are 3 commodities and 2 consumers:

$$c^1 = (1,2,3), \qquad w^1 = (1/2,1/2,1/2),$$
$$c^2 = (3,2,1), \qquad w^2 = (1/2,1/2,1/2).$$

We need c^1 and c^2 only up to positive multipliers:

$$c^1 \sim (1/6,2/6,3/6), \quad c^2 \sim (3/6,2/6,1/6).$$

Thus, c^1 and c^2 can be considered as points of the unit price simplex σ.

Figure 2 illustrates the polyhedral complexes of the model. Each of both complexes has 17 cells. **Figure 3** illustrates the arising complementarity problem. The point c^{12} is its solution: $c^{12} \in \Omega_{12}$. Thus, the corresponding vector $p^* = (3/8,2/8,3/8)$ is the equilibrium price vector of the model.

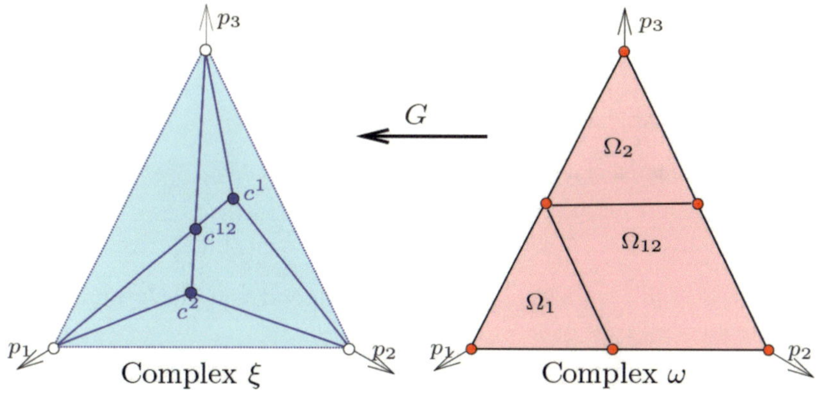

Figure 2. Polyhedral complexes in exchange model.

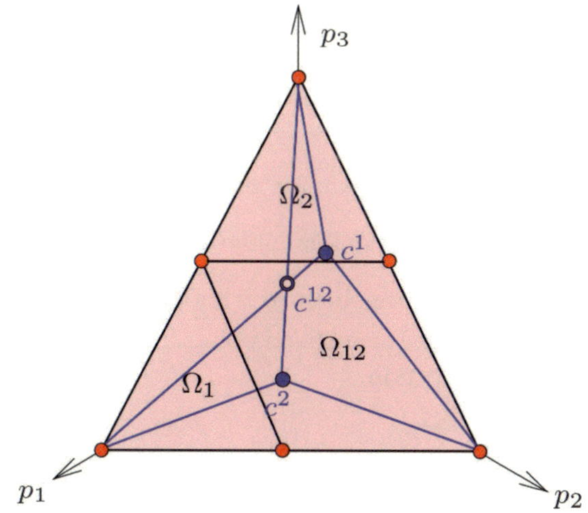

Figure 3. Complementarity problem: c^{12} is the solution.

5. The Fisher's model

1°. Reduction to optimization problem

A special class of the models is formed by the models with *fixed budgets*. This is the case when each consumer has all commodities in equal quantities: $w_j^i = \lambda_i$ for all $j \in J$, and thus, $(p, w^i) = \lambda_i$ for all $p \in \sigma$. Such a model is known as the Fisher's model. Note that we have this case in the abovementioned example.

The main feature of these models is the *potentiality* of the mappings G associated with the arising polyhedral complementarity problems.

Let f be the function on $\mathbb{R}^n$ that $f(p)$ for $p \in \sigma$ is the optimal value in the transportation problem of the model, and $f(p) = -\infty$ for $p \notin \sigma$. This function is piecewise linear and concave. It is natural to define its subdifferential using the subdifferential of convex function $(-f)$: $\partial f(p) = -\partial(-f)(p)$.

Let G be the mentioned associated mapping.

Theorem 1. *The subdifferential of the function f has the representation:*

$$\partial f(p) = \{ \ln q + te | q \in G(p), t \in \mathbb{R}\},$$

where $e = (1, \ldots, 1)$ and $\ln q = (\ln q_1, \ldots, \ln q_n)$. (The addend te in this formula arises because it holds $\sum_{j \in J} p_j = 1$ for $p \in \sigma$.)

Consider the convex function h, defining it as follows:

$$h(p) = \left\{ \begin{array}{rl} (p, \ln p), & \text{for } p \in \sigma^\circ, \\ 0, & \text{for } p \in \partial\sigma, \\ -\infty, & \text{for } p \notin \sigma. \end{array} \right\}$$

Introduce the function

$$\varphi(p) = h(p) - f(p) \tag{4}$$

Theorem 2. *The fixed point of G coincides with the minimum point of the convex function $\varphi(p)$ on σ°.*

Another theorem for the problem can be obtained if we take into account that the mapping G and the inverse mapping G^{-1} have the same fixed points. For the introduced concave function f, we can consider the conjugate function:

$$f^*(y) = \inf_z \{(y, z) - f(z)\}$$

(see [14]) With this function, we associate the function $\psi(q) = f^*(\ln q)$, which is defined on σ°.

Proposition 1. *For the Fisher's model, the following formula is valid*:

$$f^*(\ln q) = -\sum_{i\in I} \lambda_i \max_{j\in J} \ln \frac{c_j^i}{q_j} \tag{5}$$

Theorem 3. *The fixed point of G is the maximum point of the concave function $\psi(q)$ on σ°.*

For the functions $\varphi(p)$ and $\psi(q)$, there is a duality relation as for dual programs of linear programming:

Proposition 2. *For all $p, q \in \sigma^\circ$ the inequality*

$$\varphi(p) \geq \psi(q)$$

holds. This inequality turns into equality only if $p = q$.

Corollary. $\varphi(r) = \psi(r)$ *if and only if the point r is the fixed point of the mapping G.*

Thus, the equilibrium problem for the Fisher's model is reduced to the optimization one on the price simplex. It should be noted that this reduction is different from well-known one given by Eisenberg and Gale [6].

2°. Algorithms

The mentioned theorems allow us to propose two finite algorithms for searching fixed points.

Algorithmically, they are based on the ideas of suboptimization [11], which were used for minimization quasiconvex functions on a polyhedron. In considered case, we exploit the fact that the complexes ω and ξ define the cells structure on σ° similarly to the faces structure of a polyhedron.

For implementation of the algorithms, we need to get the optimum point of the function $\varphi(p)$ or $\psi(q)$ on the affine hull of the current cell.

Consider a couple of two cells $\Omega \in \omega$, $\Xi \in \xi$ corresponding to each other.

Let $L \supset \Omega$, $M \supset \Xi$ be their affine hulls. It will be shown that $L \cap M$ is singleton.

Let be $\{r\} = L \cap M$.

Lemma. *The point r is the minimum point of the function $\varphi(p)$ on L and the maximum point of the function $\psi(q)$ on M.*

Now, we describe the general scheme of the algorithm [8] that is based on Theorem 2. The other one using the Theorem 1 is quite similar [9].

On the current k-step of the process, there is a structure $\mathcal{B}_k \in \mathfrak{B}$. We consider the cells $\Omega_k = \Omega(\mathcal{B}_k)$, $\Xi_k = \Xi(\mathcal{B}_k)$ and have the point $q^k \in \Xi_k$. Let $L_k \supset \Omega_k$, $M_k \supset \Xi_k$ be the affine hulls of these cells. We need to obtain the point of their intersection r^k.

Return to the transportation problem of the model and to the descriptions of cells. Consider the graph $\Gamma(\mathcal{B}_k)$. This graph can have more than one connected components. Let τ be number of connected components, and $i \in I_\nu$, $(m+j)$ for $j \in J_\nu$ be the vertices of ν-th component. It is easy to verify that the linear system (3) for L_k is going to be equivalent to this one:

$$\sum_{j \in J_\nu} p_j = \sum_{i \in I_\nu} \lambda_i, \qquad \nu = 1, \ldots, \tau. \tag{6}$$

The linear system (1) for the cell Ξ_k defines coordinates q_j on each connected component up to a positive multiplier:

$$q_j = t_\nu q_j^k, \qquad j \in J_\nu.$$

To obtain the coordinates of the point r^k, we need to put $p_j = q_j$ in corresponding Eq. (6), which gives the multiplier t_ν.

For the obtained point, r^k can be realized in two cases.

(i) $r^k \notin \Xi_k$. Since r^k is a maximum point on M_k for the strictly concave function $\psi(q)$, the value of the function will increase for the moving point $q(t) = (1-t)q^k + tr^k)$ when t increases in [0,1]. In considered case, this point reaches a face of Ξ_k at some $t = t^* < 1$. Some of corresponding inequalities (2) for $p = q(t^*)$ is fulfilled as equality. Choose one of them. Corresponding edge $(l, m+j)$ will be added to graph. It unites two of connected components. We obtain $\mathcal{B}_{k+1} = \mathcal{B}_k \cup \{(l,j)\}$, accept $q^{k+1} = q(t^*)$ and pass to the next step.

It should be noted that the dimension of the cell Ξ reduces. It will certainly be $r^k \in \Xi_k$ when the current cell Ξ_k degenerates into a point, and we have $r^k = q^k$. But it can occur earlier.

(ii) $r^k \in \Xi_k$. In this case, we can assume $q^k = r^k$. Otherwise, we can simply replace q^k by r^k with an increase of the function's $\psi(q)$ value. We verify $q^k \in \Omega_k$? For this, we obtain from the equations of the transportation problem the variables z_{ij}, $(i,j) \in \mathcal{B}_k$, as linear functions $z_{ij}(p)$ and check $z_{ij}(q^k) \geq 0$. If it is true, the point q^k is the required fixed point. Otherwise, we have $z_{sl}(q^k) < 0$. We accept $\mathcal{B}_{k+1} = \mathcal{B}_k \ \{(s,l)\}$, $q^{k+1} = q^k$ and pass to the next step.

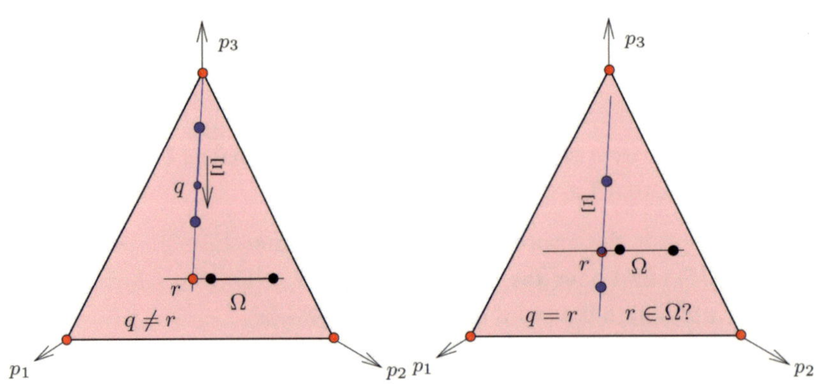

Figure 4. Illustration of one step of the algorithm.

Theorem 4. *If the transportation problem of the model is dually nondegenerate, the described suboptimization method leads to an equilibrium price vector in a finite number of steps.*

Figure 4 illustrates two described cases on one step of the algorithm. The point $q \in \Xi$ is the current point of the step.

6. Illustrative example

We show how the described method works on the Fisher's model example of Section 3.

For the start, we need a structure $\mathcal{B}^1 \in \mathfrak{B}$ and a point $q^1 \in \Xi(\mathcal{B}^1)$. We depict the structures as matrices $m \times n$ with elements from $\{\times, \cdot\}$, and $\times$ corresponds to an element of $\mathcal{B}$. For example, the structure $\mathcal{B}_{12} = \{(1,2),(1,3),(2,1),(2,2)\}$ will be depicted as the matrix

$$\mathcal{B}_{12} = \begin{pmatrix} \cdot & \times & \times \\ \times & \times & \cdot \end{pmatrix}$$

(this is the structure for the cell Ω_{12}). Let us start with the structure

$$\mathcal{B}^1 = \begin{pmatrix} \times & \cdot & \cdot \\ \times & \cdot & \cdot \end{pmatrix}$$

It means that both consumers prefer only first good. Let us choose as q^1 the price vector $q^1 = (0.05, 0.35, 0.6)$. It is easy to verify that $\mathcal{B}^1 \in \mathfrak{B}$ and $q^1 \in \Xi(\mathcal{B}^1)$.

Step 1. The graph $\Gamma(\mathcal{B}^1)$ has three connected components and the system (6) has the form

$$p_1 = 1, p_2 = 0, p_3 = 0.$$

Thus, we have $r^1 = (1,0,0)$. The cell $\Xi(\mathcal{B}^1)$ is given by the system

$$\frac{q_1}{1} \le \frac{q_2}{2}, \tag{7}$$

$$\frac{q_1}{1} \le \frac{q_3}{3}, \tag{8}$$

We have $q^1 \in \Xi(\mathcal{B}^1)$ and $r^1 \notin \Xi(\mathcal{B}^1)$. It is the case (i) in the description of algorithm. We have to move the point q^1 to the point r^1. For the moving point $q(t)$ it will be:

$$q_1(t) = 0.05 + 0.95t, \quad q_2(t) = (1-t)0.35, \quad q_3(t) = (1-t)0.6.$$

This point reaches a face of $\Xi(\mathcal{B}^1)$ at $t = t^* = 0.1111$: the inequality (7) for $q = q(t^*)$ is fulfilled as equality. We obtain $\mathcal{B}^2 = \mathcal{B}^1 \cup \{(1,2)\}$ and $q^2 = q(t^*)$.

$$\mathcal{B}^2 = \begin{pmatrix} \times & \times & \cdot \\ \times & \cdot & \cdot \end{pmatrix}$$
$$q^2 = (0.1556, 0.3111, 0.5333).$$

Step 2. The graph $\Gamma(\mathcal{B}^2)$ has two connected components and the system (6) has the form

$$p_1 + p_2 = 1, p_3 = 0.$$

For the point r^2, we have to consider this system with the additional equation

$$\frac{p_1}{1} = \frac{p_2}{2},$$

corresponding to (7), this gives $r^2 = (0.3333, 0.6667, 0)$. We have $r^2 \notin \Xi(\mathcal{B}^2)$ since the inequality (8) is violated. The new moving point $q(t) = (1-t)q^2 + tr^2$ has the coordinates:

$$q_1(t) = 0.1556 + 0.1777t, \quad q_2(t) = 0.3111 + 0.3556t, \quad q_3(t) = 0.5333 - 0.5333t.$$

At $t^* = 0.0625$ this point reaches the boundary of the cell $\Xi(\mathcal{B}^2)$. It is the point c^1 in the simplex σ. The inequality (8) for $q = q(t^*)$ is fulfilled as equality. Thus, we obtain:

$$\mathcal{B}^3 = \begin{pmatrix} \times & \times & \times \\ \times & \cdot & \cdot \end{pmatrix}$$
$$q^3 = (0.1667, 0.3333, 0.5).$$

Step 3. Now, we have the case (ii) in the description of algorithm: the cell $\Xi(\mathcal{B}^3)$ contains unique point q^3 and thus $r^3 = q^3$. We have to verify $r^3 \in \Omega(\mathcal{B}^3)$? For this, we obtain from the equations of the transportation problem the variables z_{ij}, $(i,j) \in \mathcal{B}^3$, and check $z_{ij}(q^3) \geq 0$. For these variables, we have the system:

$$z_{12} = q_2^3, \quad z_{13} = q_3^3, \quad z_{21} = 0.5, \quad z_{11} = q_1^3 - 0.5$$

We obtain $z_{11} = 0.1667 - 0.5 = -0.3333 < 0$. Thus the element $(1,1)$ should be removed from the structure $\mathcal{B}^3$:

$$\mathcal{B}^4 = \begin{pmatrix} \cdot & \times & \times \\ \times & \cdot & \cdot \end{pmatrix}$$
$$q^4 = q^3.$$

Step 4. We have to obtain the point r^4. The graph $\Gamma(\mathcal{B}^4)$ has two connected components and the system for this point has the form:

$$p_1 = 0.5, \quad p_2 + p_3 = 0.5,$$
$$\frac{p_2}{2} = \frac{p_3}{3}.$$

Hence,

$$r^4 = (0.5, 0.2, 0.3).$$

It is easy to see that the description of the cell $\Xi(\mathcal{B}^4)$ has the form:

$$\begin{aligned} &\frac{q_2}{2} = \frac{q_3}{3}, \quad \frac{q_1}{1} \geq \frac{q_3}{3}, \\ &\frac{q_1}{3} \geq \frac{q_2}{2}. \end{aligned} \tag{9}$$

For $q = r^4$, the inequality (9) is violated, so $r^4 \notin \Xi(\mathcal{B}^4)$. For the new moving point $q(t)$ we have:

$$q_1(t) = 0.1667 + 0.3333t, \quad q_2(t) = 0.3333 - 0.1333t, \quad q_3 = 0.5 - 0.2t.$$

At $t^* = 0.625$ the inequality (9) becomes equality, the point $q(t)$ attains to the point $c^{12} = (0.375, 0.25, 0.375)$ that is the boundary of $\Xi(\mathcal{B}^4)$. We obtain the new structure:

$$\mathcal{B}^5 = \begin{pmatrix} \cdot & \times & \times \\ \times & \times & \cdot \end{pmatrix}$$

and new point $q^5 = c^{12}$. It is easy to verify that we obtain the equilibrium of the model.

The equilibrium price vector is

$$\tilde{p} = (0.375, 0.25, 0.375)$$

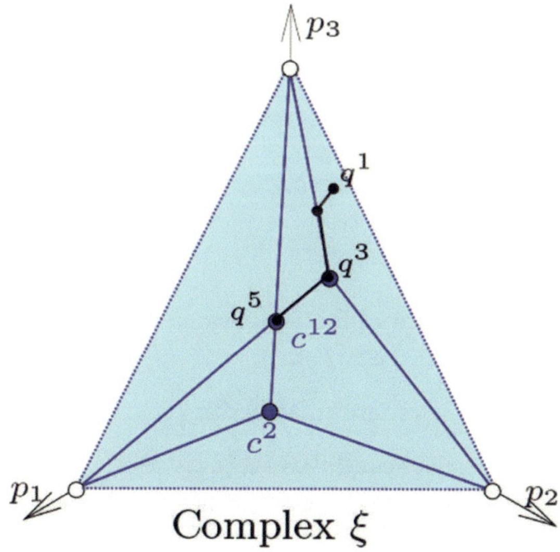

Figure 5. Movement to equilibrium in the example model.

The optimal solutions of the consumer's problems are:

$$\tilde{x}^1 = (0, 0.5, 1), \quad \tilde{x}^2 = (1, 0.5, 0).$$

Figure 5 shows the moving of the point $q(t)$ to the equilibrium.

7. Method of meeting paths

The described algorithms are nonapplicable for the general linear exchange model, when the budgets of consumers are not fixed. In this case, the associating mapping G no longer has the property of potentiality. But, the complementarity approach makes possible to propose a modification of the proses [3]. We name it *a method of meeting paths*.

As mentioned earlier, on the current k-step of the process, we have a structure $\mathcal{B}_k \in \mathfrak{B}$. We consider two cells $\Omega_k = \Omega(\mathcal{B}_k)$, $\Xi_k = \Xi(\mathcal{B}_k)$ and two points $p^k \in \Omega_k$, $q^k \in \Xi_k$. Let $L_k \supset \Omega_k$, $M_k \supset \Xi_k$ be the affine hulls of these cells. For the points of their intersection $L_k \cap M_k$, we obtain from (1), (3) the common system:

$$\frac{p_k}{c_k^i} = \frac{p_j}{c_j^i} \qquad (i,k), (i,j) \in B_k, \tag{10}$$

$$\sum_{j \in J_\nu} p_j = \sum_{i \in I_\nu} (p, w^i), \qquad \nu = 1, \ldots, \tau, \tag{11}$$

where the sets J_ν, J_ν correspond to ν-th connected component of the graph $\Gamma(\mathcal{B}_k)$.

Under some assumption about starting structure this system has rank $(n-1)$ and under additional condition $\sum_{j \in J} p_j = 1$ the system defines uniquely the solution $r^k = r(\mathcal{B}_k)$. This is the intersection point of the affine hulls of the cells $\Omega(\mathcal{B})$ and $\Xi(\mathcal{B})$. It can be shown that $r^k \in \sigma$.

If $r^k \in \Omega(\mathcal{B}_k)$, $r^k \in \Xi(\mathcal{B}_k)$, we have an equilibrium price vector.

Otherwise, we consider for $t \in [0, 1)$ two moving points:

$$p(t) = p^k + t(r^k - p^k), \quad q(t) = q^k + t(r^k - q^k)$$

It can be shown that in consequence of the assumption $c^i > 0$, $\forall i \in I$, there exists $t^* = \max t$ under the conditions $p(t) \in \Omega(\mathcal{B}_k)$, $q(t) \in \Xi(\mathcal{B}_k)$.

It is the case when $t^* < 1$. The two variants may occur:

(i) t^* is limited by some of the inequalities $z_{ij}(p(t)) \geq 0$, $(i,j) \in \mathcal{B}_k$. Corresponding pair (i,j) should be removed from $\mathcal{B}_k$: $\mathcal{B}_{k+1} = \mathcal{B}_k \setminus \{(i,j)\}$. We accept $q^{k+1} = q(t^*)$, $p^{k+1} = p(t^*)$ and pass to the next step.

(ii) t^* is limited by some of the inequalities (2) in description of the cell $\Xi(\mathcal{B}_k)$. Corresponding pair (i,l) should be added to $\mathcal{B}_k$: $\mathcal{B}_{k+1} = \mathcal{B}_k \cup \{(i,l)\}$. We accept $q^{k+1} = q(t^*)$, $p^{k+1} = p(t^*)$ and pass to the next step.

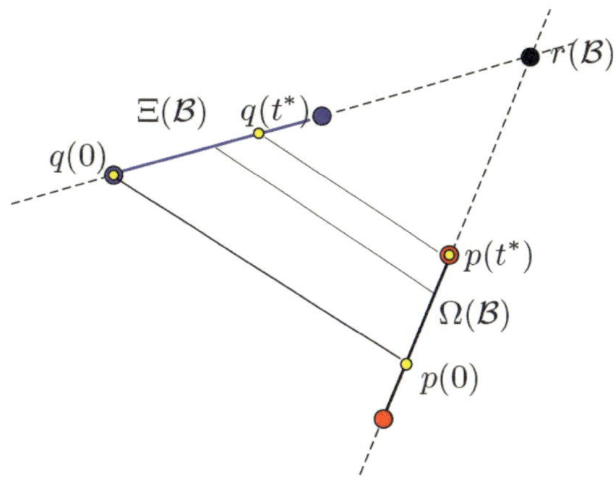

Figure 6. Illustration of a meeting paths step.

We consider the situation when t^* is limited by both above conditions as degenerate.

Nondegeneracy condition. *Only one of the above two cases can occur.*

This condition will be satisfied if a bit to move the starting points p^0, q^0.

Under this condition, it holds $t^* > 0$. To justify this, the following lemma was proved [3].

Lemma *Let A be a nonnegative and indecomposed matrix, and x is its positive eigenvector, λ is the corresponding eigenvalue. If for a positive vector $\tilde{x}$ the vector $\tilde{z} = \lambda\tilde{x} - A\tilde{x}$ has all components equal zero except $\tilde{z}_{i_1}, \tilde{z}_{i_2}$, then the following two conditions are equivalent*:

$$\tilde{z}_{i_1} \geq 0 \Longleftrightarrow \frac{\tilde{x}_{i_1}}{\tilde{x}_{i_2}} \geq \frac{x_{i_1}}{x_{i_2}}.$$

Theorem 5. *Under nondegeneracy condition, the process of meeting paths is always finite.*

Figure 6 illustrates one step of this method. In the figure, the point $p(t)$ reaches the face of its cell earlier than the point $q(t)$ does. For the next step the cell Ω will be reduced, the cell Ξ will be extended.

It should be noted that for the model with variable budgets, an iterative method was proposed [10] that uses the developed simple algorithm for Fisher's model in each step of the process.

8. Generalizations

1. The models with upper bounds: the considered approach permits to develop the algorithms for deferent variations of the classical exchange model. The simplest of those models is the model in which the costs are limited for certain goods (the spending

constraints model [7]): $p_j x_j^i \leq \beta_{ij}$. In this case, the mappings G associated with the arising polyhedral complementarity problem are potential too. Some modifications of the developed algorithms are needed. More difficult is the model with upper on the purchase volumes of goods. In this case, the mappings G are not potential and algorithm becomes more complicated. Such a model arises if the functions of participants are not linear, but piecewise linear concave separable [5].

2. The generalized linear exchange model: the polyhedral complementarity approach is applicable to models with the production sector too. Some firms are added, those supply goods to the market. Describe more in detail one of those models.

The model with n products, m participants-consumers, and l participants-firms is considered. Let $J = \{1, \dots, n\}, I = \{1, \dots, m\}$, and $K = \{m+1, \dots, m+l\}$ be the sets of the numbers of products, consumers, and firms. Thus, $S = I \cup K$ is the set of numbers of all participants.

The consumer $i \in I$ has the initial endowments $w^i \in R_+^n$ and also the initial money stock α_i. His total budget after selling the initial endowments is equal to $\alpha_i + (p, w^i)$. Thus, the i th consumer will choose the purchase vector x^i looking for an optimal solution to the following problem:

$$(c^i, x^i) \rightarrow \max$$

under the conditions

$$(p, x^i) \leq \alpha_i + (p, w^i),$$

$$x^i \geq 0.$$

The firm $k \in K$ plans to deliver to the market the products to a total sum of at least λ_k. If $x^k = (x_1^k, \dots, x_n^k)$ denotes a plan of k th firm then the total cost of such a supply at the prices p_j equals (p, x^k). The quality of the plan is estimated by the firm in tending to minimize the function $(c^k, , x^k)$. Here, $c^k = (c_1^k, \dots, c_n^k)$ is a fixed nonnegative vector whose components determine a comparative scale of the "undesirability" of various products for the firm (e.g., their relative production costs).

Thus, the k th firm makes its choice according to a solution of the optimization problem:

$$(c^k, x^k) \rightarrow \min$$

under the conditions

$$(p, x^k) \geq \lambda_k,$$
$$x^k \geq 0.$$

An equilibrium is defined by a price vector $\tilde{p}$ and a collection of vectors $\tilde{x}^i$ and $\tilde{x}^k$ $i \in I$ and $k \in K$, representing some solutions to optimization problems of the participants for $p = \tilde{p}$ and satisfying the balance of products:

$$\sum_{i\in I}\tilde{x}^i=\sum_{k\in K}\tilde{x}^k+\sum_{i\in I}w^i.$$

From this follows that we have to suppose $\sum_{i\in I}\alpha_i=\sum_{k\in K}\lambda_k$. As before, we suppose also that $\sum_{j\in J}p_j=1$ and $\sum_{i\in I}w^i=e$ with $e=(1,\dots,1)$. Thus, in the equilibrium, we have

$$\sum_{i\in I}\tilde{x}^i=\sum_{k\in K}\tilde{x}^k+e. \tag{12}$$

The polyhedral complementarity approach can be used for this generalized model as well. The main results remain valid [4], but the consideration becomes more complicated. Some features are discussed.

The structure notion is generalized:

Definition. *A set $B\subset S\times J$ is named a structure, if for each $s\in S$ there exists $(s,j)\in\mathcal{B}$.*

As before, we suppose that all vectors $c^s, s\in S$ are positive. The parametric transportation problem of the model is changed and becomes a net problem:

$$\sum_{i\in I}\sum_{j\in J}z_{ij}\ln c_j^i-\sum_{k\in K}\sum_{j\in J}z_{kj}\ln c_j^k \ \to\ \max$$

$$-\sum_{j\in J}z_{ij}=-\alpha_i-(p,w^i),\quad i\in I,$$

$$\sum_{i\in I}z_{ij}-\sum_{k\in K}z_{kj}=p_j,\quad j\in J,$$

$$\sum_{j\in J}z_{kj}=\lambda_k,\quad k\in K,$$

$$z_{ij}\ge 0,\quad z_{kj}\ge 0,\qquad i\in I,\quad j\in J,\quad k\in K.$$

It can be shown that this problem is solvable for all $p\in\sigma$.

As mentioned before, we consider the family $\mathfrak{B}$ of structures B: it is the collection of all dual feasible basic index sets of the transportation problem and of all their subsets being structures.

For each $B\in\mathfrak{B}$, we define the balance zone $\Omega(B)$ and the preference zone $\Xi(B)$. The description of these sets is quite similar to those of the classical case. Thus, in this way, we again obtain two polyhedral complexes.

Theorem 6. *A vector $\tilde{p}\in\sigma^\circ$ is an equilibrium price vector of generalized linear exchange model if and only if $\tilde{p}\in\Omega(B)\cap\Xi(B)$ for some $B\in\mathfrak{B}$.*

The generalized model can be considered with fixed budgets, and in this way, we obtain the generalization of the Fisher's model. The budget condition of the consumer i remains the same: $(p,x^i)\le\lambda_i$. But for the $\lambda_i, i\in I$ and $\lambda_k, k\in K$, we obtain the condition $\sum_{i\in I}\lambda_i=1+\sum_{k\in K}\lambda_k$.

For this variant of model, we have the reduction to optimization problems as well. To do this, we consider the function $f(p)$, which gives the optimal value of the transportation problem by given price vector p. Having this function, we introduce as before the functions $\varphi(p) = (p, \ln p) - f(p)$ and $\psi(q) = f^*(\ln q)$. For these functions, the main results of classical case remain valid.

Theorem 7. *A vector* $\tilde{p}$ *is an equilibrium price vector if and only if* $\tilde{p}$ *is a minimum point of the function* φ *on* σ°.

Theorem 8. *A vector* $\tilde{p}$ *is an equilibrium price vector if and only if* $\tilde{p}$ *is a maximum point of the function* ψ *on* σ°.

The finite algorithms developed for Fischer's model do not require any significant changes and are applicable for this generalized model.

3. The production-exchange models Arrow-Debreu type: these are modifications of previous model. Describe the simplest variant of the model. On the market, there is one unit of each good. The firms produce additional goods, spending some resource that is limited and seek to maximize revenue from the sale of manufactured goods. Thus, the k th firm solves the following problem:

$$\sum_{j \in J} p_j x_j^k \to \max$$

$$\sum_{j \in J} d_j^k x_j^k \le \zeta_k,$$

$$x_j^k \ge 0, \quad j \in J.$$

Here, ζ_k is allowable resource and d_j^k indicate the resource cost per unit of product j.

Let $\lambda_k(p)$ be the optimal value of this problem. The consumer $i \in I$ has the initial money stock α_i, $\sum_{i \in I} \alpha_i = 1$. The revenues of the firms are divided between consumers in some proportions, those are given by θ_{ik}. The total budget of i th consumers becomes $\alpha_i + \sum_{k \in K} \theta_{ik} \lambda_k(p)$. Thus, the i th consumer has the following problem:

$$(c^i, x^i) \to \max$$

under the conditions

$$(p, x^i) \le \alpha_i + \sum_{k \in K} \theta_{ik} \lambda_k(p),$$

$$x^i \ge 0.$$

The condition of good balances in equilibrium is given as before by the equality (12).

The polyhedral complementarity approach is applicable for this model too, but the consideration becomes much more complicated. An iterative method can be developed that uses the abovementioned generalized linear exchange model as an auxiliary in each step of the process.

Acknowledgements

This chapter was supported by the Russian Foundation for Basic Research, project 16-01-00108 À.

Author details

Vadim I. Shmyrev[1,2]*

*Address all correspondence to: shmyrev.vadim@mail.ru

1 Novosibirsk State University, Novosibirsk, Russia

2 Sobolev Institute of Mathematics, Russian Academy of Sciences, Novosibirsk, Russia

References

[1] Eaves BC. A finite algorithm for linear exchange model. Journal of Mathematical Economics. 1976;**3**(2):197-204

[2] Shmyrev VI. On an approach to the determination of equilibrium in elementary exchange models. Soviet Mathematics - Doklady. 1983;**27**(1):230-233

[3] Shmyrev VI. An algorithm for the search of equilibrium in the linear exchange model. Siberian Mathematical Journal. 1985;**26**:288-300

[4] Shmyrev VI. A generalized linear exchange model. Journal of Applied and Industrial Mathematics. 2008;**2**(1):125-142

[5] Shmyrev VI. An iterative approach for searching an equilibrium in piecewise linear exchange model. In: Kochetov Yu. et al., editors. Lecture Notes in Computer Sciences. vol. 9869. Heidelberg, Germany: Springer; 2016. pp. 61-72

[6] Eisenberg E, Gale D. Consensus of subjective probabilities: The pari-mutuel method. The Annals of Mathematical Statistics. 1959;**30**(1):165-168

[7] Devanur NR, Papadimitriou CH, Saberi A, Vazirani VV. Market equilibrium via a primal–dual algorithm for a convex program. Journal of the ACM (JACM). 2008;**55**(5):22

[8] Shmyrev VI. An algorithmic approach for searching an equilibrium in fixed budget exchange models. In: Driessen TS et al., editors. Russian Contributions to Game Theory and Equilibrium Theory. Berlin, Germany: Springer; 2006. pp. 217-235

[9] Shmyrev VI. An algorithm for finding equilibrium in the linear exchange model with fixed budgets. Journal of Applied and Industrial Mathematics. 2009;**3**(4):505-518

[10] Shmyrev VI, Shmyreva NV. An iterative algorithm for searching an equilibrium in the linear exchange model. Siberian Advances in Mathematics. 1996;**6**(1):87-104

[11] Rubinstein GS, Shmyrev VI. Methods for minimization quasiconvex function on polyhadron (in Russian). Optimization. 1971;**1**(18):82-117

[12] Lemke CE. Bimatrix equilibrium points and mathematical programming. Management Science. 1965;**2**(7):681-689

[13] Gale D. The linear exchange model. Journal of Mathematical Economics. 1976;**3**(2):205-209

[14] Rockafellar R. Convex Analisis. USA: Princeton University; 1970

Chapter 3

Multicriteria Support for Group Decision Making

Andrzej Łodziński

Additional information is available at the end of the chapter

http://dx.doi.org/10.5772/intechopen.79935

Abstract

This chapter presents the support method for group decision making. A group decision is when a group of people has to make one joint decision. Each member of the group has his own assessment of a joint decision. The decision making of a group decision is modeled as a multicriteria optimization problem where the respective evaluation functions are the assessment of a joint decision by each member. The interactive analysis that is based on the reference point method applied to the multicriteria problems allows to find effective solutions matching the group's preferences. Each member of the group is able to verify results of every decision. The chapter presents an example of an application of the support method in the selection of the group decision.

Keywords: multicriteria optimization problem, equitably efficient decision, scalarizing function, decision support systems

1. Introduction

The chapter presents the support method for group decision making—when a group of people who have different preferences want to make one joint decision.

The selection process of a group decision can be modeled with the use of game theory [1–3].

In this chapter, the choice of the group decision is modeled as a multicriteria problem. The individual coordinates of this optimization problem are functions to evaluate a joint decision by each person in the group. This allows one to take into account preferences of all members in the group. Decision support is an interactive process of proposals for subsequent decisions, by each member in the group and his evaluations. These proposals are parameters of the multicriteria optimization problem. The solution of this problem is assessed by members in the group. Each member can accept or refuse the solution. In the second case, a member gives his new proposal and the problem is resolved again.

2. Modeling of group decision making

The problem of choosing a group decision is as follows. There is a group of k members. There is given a set X_0—the feasible set. For each member i, $i = 1, 2, \ldots, k$, a decision evaluation function f_i is defined, which is an assessment of a joint decision. The assessment of the joint decision is to be made by all members in the group.

The problem of group decision making is modeled as multicriteria optimization problem:

$$\max_x \{(f_1(x), \ldots, f_k(x)) : \ x \in X_0\}, \tag{1}$$

where $1, 2, \ldots, k$ are particular members, $X_0 \subset R^n$ is the feasible set, $x = (x_1, x_2, \ldots, x_n) \in X_0$ is a group decision, $f = (f_1, f_2, \ldots, f_k)$ is the vector function that maps the decision space $X_0 = R^n$ into the criteria space $Y_0 = R^k$, and specific coordinates $y_i = f_i(x), \ \ i = 1, 2, .., m$ represent the scalar evaluation functions—the result of a decision $x \ i - th$ member $i = 1, 2, \ldots, k$.

The purpose of the problem (1) is to support the decision process to make a decision that will be the most satisfactory for all members in the group.

Functions $f_1, \ldots, f_k$ introduce a certain order in the set of decision variables—preference relations:

$$x^1 \succ x^2 \Leftrightarrow f_1(x^1) \geq f_2(x^2), \ldots, f_k(x^1) \geq f_k(x^2) \ \ \wedge \ \ \exists j \, f_j(x^1) > f_j(x^2). \tag{2}$$

At point x^1, all functions have values greater than or equal to the value at point x^2, and at least one is greater.

The multicriteria optimization model (1) can be rewritten in the equivalent form in the space of evaluations. Consider the following problem:

$$\max_x \{(y_1, \ldots, y_k) : \ \ y \in Y_0\}, \tag{3}$$

where $x \in X$ is a vector of decision variables, $y = (y_1, \ldots, y_k)$ is the evaluation vector and particular coordinates y_i represent the result of a decision $x \ i - th$ member $i = 1, 2, \ldots, k$, and $Y_0 = f(X_0)$ is the set of evaluation vectors.

The vector function $y = f(x)$ assigns to each vector of decision variables x an evaluation vector $y \in Y_0$ that measures the quality of decision x from the point of view of all members in the group. The set of results achieved Y_0 is given in the implicit form—through a set of feasible decisions X_0 and the mapping of a model $f = (f_1, f_2, \ldots, f_k)$. To determine the value y, the simulation of the model is necessary: $y = f(x) \ \ for \ \ x \in X_0$.

3. Equitably efficient decision

Group decision making is modeled as a special multicriteria optimization problem—the solution should have the feature of anonymity—no distinction is made between the results that differ in the orientation coordinates and the principle of transfers. This solution of the problem is named an equitably efficient decision. It is an efficient decision that satisfies the additional property-the property of preference relation anonymity and the principle of transfers.

Nondominated solutions (optimum Pareto) are defined with the use of preference relations which answer the question: which one of the given pair of evaluation vectors $y^1, y^2 \in R^k$ is better? This is the following relation:

$$y^1 \succ y^2 \Leftrightarrow y_i^1 \geq y_i^2 \forall i = 1, \dots, m \ \wedge \ \exists j \ y_j^1 > y_j^2. \tag{4}$$

The vector of evaluation $\hat{y} \in Y_0$ is called the nondominated vector; if there is no such vector $y \in Y_0$, that $\hat{y}$ is dominated by y. Appropriate acceptable decisions are specified in the decision space. The decision $\hat{x} \in X_0$ is called efficient decision (Pareto efficient) if the corresponding vector of evaluations $\hat{y} = f(\hat{x})$ is a nondominated vector [4, 5].

In the multicriteria problem (1), which is used to make a group decision for a given set of the evaluation functions, only the set of the evaluation functions is important without taking into account which function is taking a specific value. No distinction is made between the results that differ in the arrangement. This requirement is formulated as the property of anonymity of preference relation.

The relation is called an anonymous (symmetric) relation if, for every vector $y = (y_1, y_2, \dots, y_k) \in R^k$ and for any permutation P of the set $\{1, \dots, k\}$, the following property holds:

$$\left(y_{P(1)}, y_{P(2)}, \dots, y_{P(k)}\right) \approx (y_1, y_2, \dots, y_k) \tag{5}$$

The relation of preferences that would satisfy the anonymity property is called symmetrical relation. Evaluation vectors having the same coordinates, but in a different order, are identified. A nondominated vector satisfying the anonymity property is called symmetrically nondominated vector.

Moreover, the preference model in group decision making should satisfy the principle of transfers. This principle states that the transfer of small amount from an evaluation vector to any relatively worse evaluation vector results in a more preferred evaluation vector. The relation of preferences satisfies the principle of transfers, if the following condition is satisfied:

for the evaluation vector $y = (y_1, y_2, \dots, y_k) \in R^k$:

$$y_{i'} > y_{i''} \Rightarrow y - \varepsilon \cdot e_{i'} + \varepsilon \cdot e_{i''} \succ y \ \text{ for } 0 < y_{i'} - y_{i''} < \varepsilon \tag{6}$$

Equalizing transfer is a slight deterioration of a better coordinate of evaluation vector and, simultaneously, improvement of a poorer coordinate. The resulting evaluation vector is strictly preferred in comparison to the initial evaluation vector. This is a structure of equalizing—the evaluation vector with less diversity of coordinates is preferred in relation to the vector with the same sum of coordinates, but with their greater diversity.

A nondominated vector satisfying the anonymity property and the principle of transfers is called equitably nondominated vector. The set of equitably nondominated vectors is denoted by $\widehat{Y}_{0E}$. In the decision space, the equitably efficient decisions are specified. The decision $\widehat{x} \in X_0$ is called an equitably efficient decision, if the corresponding evaluation vector $\widehat{y} = f(\widehat{x})$ is an equitably nondominated vector. The set of equitably efficient decisions is denoted by $\widehat{X}_{0E}$ [2, 6, 7].

Equitable dominance can be expressed as the relation of inequality for cumulative, ordered evaluation vectors. This relation can be determined with the use of mapping $\overline{T} : R^k \to R^k$ that cumulates nonincreasing coordinates of evaluation vector.

The transformation $\overline{T} : R^k \to R^k$ is defined as follows:

$$\overline{T}_i(y) = \sum_{l=1}^{i} T_i(y) \quad \text{for } i = 1, 2, \ldots, k. \tag{7}$$

Define by $T(y)$ the vector with nonincreasing ordered coordinates of the vector y, i.e. $T(y) = (T_1(y), T_2(y), \ldots, T_k(y))$, where $T_1(y) \leq T_2(y) \leq \ldots \leq T_k(y)$ and there is a permutation P of the set $\{1, \ldots, k\}$, such that $T_i(y) = y_{P(i)}$ for $i = 1, .., k$.

The relation of equitable domination $\succ_e$ is a simple vector domination for evaluation vectors with cumulated nonincreasing coordinates of evaluation vector [6, 7].

The evaluation vector y^1 equitably dominates the vector y^2 if the following condition is satisfied:

$$y^1 \succ_e y^2 \Leftrightarrow \overline{T}(y^1) \geq \overline{T}(y^2) \tag{8}$$

The solution of choosing a group decision is to find the equitably efficient decision that best reflects the preferences of all members in the group.

4. Technique of generating equitably efficient decisions

Equitably efficient decisions for a multiple criteria problem (1) are obtained by solving a special problem in multicriteria optimization—a problem with the vector function of the cumulative, evaluation vectors arranged in a nonincreasing order. This is the following problem.

$$\max_y \left\{ \left(\overline{T}_1(y), \overline{T}_2(y), \ldots, \overline{T}_k(y) \right) : \quad y \in Y_0 \tag{9}$$

where

http://dx.doi.org/10.5772/intechopen.79935

$y = (y_1, y_2, \ldots, y_k)$ is the evaluation vector, $\overline{T}(y) = (\overline{T}_1(y), \overline{T}_2(y), \ldots, \overline{T}_k(y))$ is the cumulative, ordered evaluation vector, and Y_0 is the set of achievable evaluation vectors.

The efficient solution of multicriteria optimization problem (9) is an equitably efficient solution of the multicriteria problem (1).

To determine the solution of a multicriteria problem (9), the scalaring of this problem with the scalaring function $s : Y_0 \times \Omega \rightarrow R^1$ is solved:

$$\max_x \{s(y, \overline{y}) : x \in X_o, \tag{10}$$

where $y = (y_1, y_2, \ldots, y_k)$ is the evaluation vector and $\overline{y} = (\overline{y}_1, \overline{y}_2, \ldots, \overline{y}_k)$ is the control parameter for individual evaluations.

It is the problem of single-objective optimization with specially created scalaring function of two variables—the evaluation vector $y \in Y$ and control parameter $\overline{y} \in \Omega \subset R^k$; we have thus $s : Y_0 \times \Omega \rightarrow R^1$. The parameter $\overline{y} = (\overline{y}_1, \overline{y}_2, \ldots, \overline{y}_k)$ is available to each member in the group that allows any member to review the set of equitably efficient solutions.

Complete and sufficient parameterization of the set of equitably efficient decision $\widehat{X}_{0E}$ can be achieved, using the method of the reference point for the problem (9). In this method the aspiration levels are applied as control parameters. Aspiration level is the value of the evaluation function that satisfies a given member.

The scalaring function defined in the method of reference point is as follows:

$$s(y, \overline{y}) = \min_{1 \le i \le k} \left(\overline{T}_i(y) - \overline{T}_i(\overline{y})_i\right) + \varepsilon \cdot \sum_{i=1}^{k} \left(\overline{T}_i(y) - \overline{T}_i(\overline{y})_i\right), \tag{11}$$

where $y = (y_1, y_2, \ldots, y_k)$ is the evaluation vector; $\overline{T}(y) = (\overline{T}_1(y), \overline{T}_2(y), \ldots, \overline{T}_k(y))$ is the cumulative, ordered evaluation vector; $\overline{y} = (\overline{y}_1, \overline{y}_2, \ldots, \overline{y}_k)$ is the vector of aspiration levels; $T(\overline{y}) = (T_1(\overline{y}), T_2(\overline{y}), \ldots, T_k(\overline{y}))$ is the cumulative, ordered vector of aspiration levels; and ε is the arbitrary, small, positive adjustment parameter.

This function is called a function of achievement. Maximizing this function with respect to y determines equitably nondominated vectors $\widehat{y}$ and the equitably efficient decision $\widehat{x}$. For any aspiration levels $\overline{y}$, each maximal point $\widehat{y}$ of this function is an equitably nondominated solution. Note, the equitably efficient solution $\widehat{x}$ depends on the aspiration levels $\overline{y}$. If the aspiration levels $\overline{y}$ are too high, then the maximum of this function is smaller than zero. If the aspiration levels $\overline{y}$ are too low, then the maximum of this function is larger than zero. This is the information for the group, whether a given aspiration level is reachable or not [4, 8].

A tool for searching the set of solutions is the function (11). Maximum of this function depends on the parameter $\overline{y}$, which is used by the members of the group to select a solution. The method for supporting selection of group decisions is as follows:

- Calculations—giving other equitably efficient decisions
- Interaction with the system—dialog with the members of the group, which is a source of additional information about the preferences of the group

The method of selecting group decision is presented in **Figure 1**.

The computer will not replace members of the group in the decision-making process; the whole process of selecting a decision is guided by all members in the group.

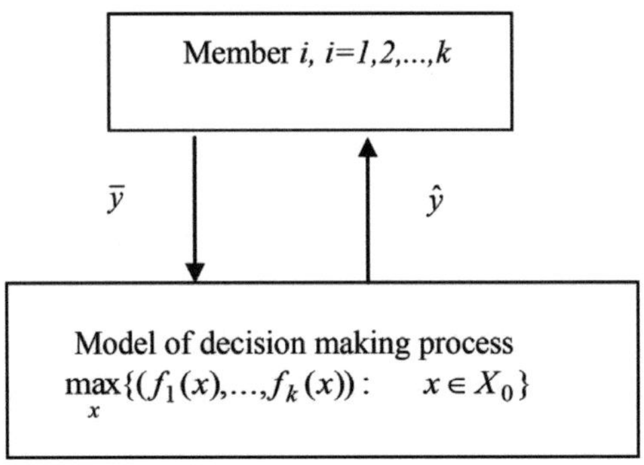

Figure 1. The method of selecting group decision.

5. Example

To illustrate the process of supporting group decision making, the following example is presented—selection of group decision by three members [8].

The problem of selecting the decision is the following:

1, 2, 3 are the members in the group.

$X_0 = x \in R^2: \ x_1 + 5 \cdot x_2 \leq 75, \ 3 \cdot x_1 + 5 \cdot x_2 \leq 95, x_1 + x_2 \leq 25, \ 5 \cdot x_1 + 2 \cdot x_2 \leq 110, \ x_1 \geq 0, x_2 \geq 0$ is the feasible set.

$x = (x_1, x_2) \in X_0$ is a group decision, belonging to the feasible set.

$f_1(x) = 10 \cdot x_1 + 60 \cdot x_2$ is the function of decision evaluation x by member 1.

$f_1(x) = 40 \cdot x_1 + 60 \cdot x_2$ is the function of decision evaluation x by member 2.

$f_1(x) = 60 \cdot x_1 + 20 \cdot x_2$ is the function of decision evaluation x by member 3.

The problem of selection of group decision is expressed in the form of multicriteria optimization problem with three evaluation functions:

$$\max_x \{(10 \cdot x_1 + 60 \cdot x_2, 40 \cdot x_1 + 60 \cdot x_2, 60 \cdot x_1 + 20 \cdot x_2) \ x \in X_0\}, \tag{12}$$

where X_0 is the feasible set and $x = (x_1, x_2) \in X_0$ is a group decision.

A solution which is as satisfying as possible for all members in the group is searched for. All members in the problem of decision making in a group should be treated in the same way, no member should be favored. The decision-making model should have the anonymity properties of preference relation and satisfy the principle of transfers. The solution of the problem should be an equitably efficient decision of the problem (12).

For solving the problem (12) the method of reference point is used.

At the beginning of the analysis, a separate single-criterion optimization is carried out for each member in the group. In this way, the best results for each member are obtained separately. This is a utopia point of the multicriteria optimization problem. This also gives information about the conflict of evaluations of group members in the decision-making problem [9, 10].

When analyzing **Table 1**, it might be observed that the big selection possibilities have members 2 and 3 and lower member 1.

For each iteration, the price of fairness (POF) for each member is calculated [4]. It is the quotient of the difference between the utopia value of a solution and the value from the solution of the multicriteria problem, in relation to the utopia value.

$$POF = \frac{yiu - \widehat{y}i}{yiu}, i = 1, 2, 3, \tag{13}$$

where yiu is the utopia value of a member i, $i = 1, 2, 3$, and yiu is the value from the solution of the multicriteria problems of a member, $i\ i = 1, 2, 3$.

The value of the POFs is a number between 0 and 1. POF values closer to zero are preferred by the members, as the solution is closer to a utopia solution. The more the values of the POFs of the members get closer to each other, the better the solution.

People in the group do control the process by means of aspiration levels. The multicriteria analysis is presented in **Table 2**.

At the beginning of the analysis (Iteration 1), members in the group define their preferences as aspiration levels equal to the values of utopia. The obtained effective leveling solution is ideal for member 2, while member 1 and member 3 would like to correct their solutions. In the next iteration, all members reduce their levels of aspiration. As a result (Iteration 2), the solution for

Optimization criterion	**Solution**		
	$\widehat{y}1$	$\widehat{y}2$	$\widehat{y}3$
Member's evaluation 1 $y1$	900	900	300
Member's evaluation 2 $y2$	750	1200	1100
Member's evaluation 3 $y3$	220	880	1320
Utopia vector	900	1200	1320

Table 1. Matrix of goal realization with the utopia vector.

	Iteration	Member 1	Member 2	Member 3
		$\widehat{y}_1$	$\widehat{y}_2$	$\widehat{y}_2$
1.	Aspiration levels $\overline{y}$	900	1200	1320
	Solution $\widehat{y}$	750	1200	1100
	POF	0.166	0	0.153
2.	Aspiration levels $\overline{y}$	850	1000	1200
	Solution $\widehat{y}$	800	1192	1007
	POF	0.111	0.006	0.224
3.	Aspiration levels $\overline{y}$	850	1000	1250
	Solution $\widehat{y}$	775	1196	1053
	POF	0.138	0.003	0.189
4.	Aspiration levels $\overline{y}$	850	1000	1300
	Solution $\widehat{y}$	750	1200	1100
	POF	0.166	0	0.153
5.	Aspiration levels $\overline{y}$	850	990	1300
	Solution $\widehat{y}$	755	1199	1090
	POF	0.161	0.0006	0.160

Table 2. Interactive analysis of seeking a solution.

member 1 has improved, while the solution for member 2 and member 3 has deteriorated. The group now wishes to correct the solution for member 3 and increases the aspiration level for member 3, but does not change the aspiration levels for members 1 and 2. As a result (Iteration 3), the solution for member 2 and member 3 has improved, while the solution for member 1 has deteriorated. The group still wishes to correct the solution for member 3 and provides a higher value of the aspiration level for member 3, but does not change the aspiration levels for members 1 and 2. As a result (Iteration 4), the solution for member 2 and member 3 has improved, but the solution for member 1 has deteriorated. The group now wishes to correct the solution for member 1 and member 3 and reduces the aspiration level for member 2, but does not reduce the aspiration levels for members 1 and 3. As a result (Iteration 5), the solution for member 1 has improved, while the solution for members 2 and 3 has deteriorated. A further change to the value of the aspiration levels causes either an improvement in the solution for member 1 and at the same time a deterioration in the solution for member 3 or vice versa, as well as slight changes in the solution for member 2. Such a solution results from the specific nature of the examined problem—the solution for member 2 lies between solutions for members 1 and 3. The group decision for Iteration 5 is as follows: $x^5 = (14.81, \quad 10.12)$.

The final choice of a specific solution depends on the preferences of the members in the group. This example shows that the presented method allows the members to get to know their decision-making possibilities within interactive analysis and to search for a solution that would be satisfactory for the group.

6. Summary

The chapter presents the method of supporting group decision making. The choice is made by solving the problem of multicriteria optimization.

The decision support process is not a one-step act, but an iterative process, and it proceeds as follows:

- Each member of the group participates in the decision-making process.
- Then, each member determines the aspiration levels for particular results of decisions. These aspiration levels are determined adaptively in the learning process.
- The decision choice is not a single optimization act, but a dynamic process of searching for solutions in which each member may change his preferences.
- This process ends when the group finds a decision that makes it possible to achieve results meeting the member's aspirations or closest to these aspirations in a sense.

This method allows the group to verify the effects of each decision and helps find the decision which is the best for their aspiration levels. This procedure does not replace the group in decision-making process. The whole decision-making process is controlled by all the members in the group.

Author details

Andrzej Łodziński

Address all correspondence to: andrzej_lodzinski@sggw.pl

Faculty of Applied Informatics and Mathematics, Warsaw University of Life Sciences, Warsaw, Poland

References

[1] Luce D, Raiffa H. Games and Decisions. Warsaw: PWN; 1966. (in Polish)

[2] Malawski M, Wieczorek A, Sosnowska H. Competition and cooperation. Game Theory in Economics and the Social Sciences. Warsaw: PWN; 1997. (in Polish)

[3] Straffin PhD. Game Theory. Warsaw: Scolar; 2004. (in Polish)

[4] Lewandowski A, Wierzbicki A, editors. Aspiration Based Decision Support Systems. Lecture Notes in Economics and Mathematical Systems. Vol. 331; Berlin-Heidelberg: Springer-Verlag; 1989

[5] Wierzbicki AP. A mathematical basis for satisficing decision making. Mathematical Modelling. 1982;**3**:391-405

[6] Łodziński A. Avoiding risk in decision making process (in Polish). Zarządzanie Przedsiębiorstwem – Teoria I Praktyka. Kraków: Wydawnictwa AGH; 2014. pp. 241-251

[7] Ogryczak W. Multicriteria optimization and decisions under risk. Control and Cybernetics. 2002;**31**(4):975-1003

[8] Wierzbicki A, Makowski N, Wessels J. Model_Based Decision Support Methodology with Environmental Applications. Laxenburg, Dordrecht: IIASA Kluwer; 2000

[9] Kostreva M, Ogryczak W, Wierzbicki A. Equitable aggregation and multiple criteria analysis. European Journal of Operational Research. 2004;**158**:362-377

[10] Krawczyk S. Mathematical Analysis of the Situation of Decision-Making. Warsaw: PWE Warsaw; 1990. (in Polish)

Chapter 4

On Non-Linearity and Convergence in Non-Linear Least Squares

Orhan Kurt

Additional information is available at the end of the chapter

http://dx.doi.org/10.5772/intechopen.76313

Abstract

To interpret and explain the mechanism of an engineering problem, the redundant observations are carried out by scientists and engineers. The functional relationships between the observations and parameters defining the model are generally nonlinear. Those relationships are constituted by a nonlinear equation system. The equations of the system are not solved without using linearization of them on the computer. If the linearized equations are consistent, the solution of the system is ensured for a probably global minimum quickly by any approximated values of the parameters in the least squares (LS). Otherwise, namely an inconsistent case, the convergence of the solution needs to be well-determined approximate values for the global minimum solution even if in LS. A numerical example for 3D space fixes coordinates of an artificial global navigation satellite system (GNSS) satellite modeled by a simple combination of first-degree polynomial and first-order trigonometric functions will be given. It will be shown by the real example that the convergence of the solution depends on the approximated values of the model parameters.

Keywords: nonlinear equation system, objective function, least squares, convergence, consistency

1. Introduction

There are two main computing classes, these are hard and soft computing. Scientists and engineers generally prefer the first-class computing because they can easily establish an explicit mathematical relationship between the model parameters and their data (observations), not in the second class. The relationships between the parameters and data can be linear or nonlinear.

If the relations are nonlinear, they should be linearized via Taylor expansion [1–7]. Therefore, the linear models can be solved by linear algebra [8–15].

To overcome complicated real-life problems whose mathematical models are not known, the soft computing techniques have been developed in the last decades. We can count well-known techniques, some as artificial neural network (ANN), artificial intelligence (AI), machine learning (ML), deep learning (DP), fuzzy logic (FL) and genetic algorithms (GA) [16–18]. The techniques inspired by the human intelligence and learning processes can be very time-consuming according to the data given in run due to their processing based on the trial-and-error method. If these techniques are roughly defined, data (experimental outcomes and observations) are separated into two parts in them, learning (or training) data and test data. Mathematical (functional and/or stochastic) relations between data and model parameters are learned from the learning data. The handled model is tested by means of the test data. After that, the trained and developed model, if meets expectations, is used to estimate for producing unobserved data for the scientific (or engineering) problems [16–18].

In the soft computing techniques, the linear algebra is also a very effective tool to solve the problem as in the hard computing ones. For this reason, we should take a short overview on linear algebra used in science and engineering [16–18].

2. Linear algebra and objective functions

Linear algebra has two basic problems. A solution of linear equations system is one of them; the other is the eigendecomposition. In this chapter, we will use both of them upon a linear equation system as a combined form (Eqs. (8)–(11)) in which we will solve the linear equations system by means of the singular value decomposition related with the eigendecomposition (or the matrix diagonalization) [8, 9, 13, 14].

Suppose an estimated unknown vector $\widehat{\mathbf{x}}_u = \mathbf{x} + \widehat{\boldsymbol{\delta}}$ (in interested model) and an experimental data (or observations which are stochastic variables) vector $\mathbf{y}_n = \widehat{\mathbf{y}} - \widehat{\boldsymbol{\varepsilon}}$ [in which an estimated data and error (residual) vectors are in order of $\widehat{\mathbf{y}}$ and $\widehat{\boldsymbol{\varepsilon}}$] by an objective function and their covariance matrices $\boldsymbol{\Sigma}_{\widehat{\mathbf{x}}} = \boldsymbol{\Sigma}_{\mathbf{x}} = \widehat{\sigma}_0^2\ \mathbf{Q}_{\mathbf{x}}$ (for the unknowns) and $\boldsymbol{\Sigma}_{\mathbf{y}} = \sigma_0^2\ \mathbf{P}^{-1}$ (for the data), respectively, with a priori variance σ_0^2 and a posteriori variance $\widehat{\sigma}_0^2$. Note that $\widehat{\mathbf{x}}$ is a non-stochastic vector before estimation, where an approximated values vector is $\mathbf{x}$ for $\widehat{\mathbf{x}}$ (hat-sign "^" shows an estimated value for interested parameter according to an objective function). In addition, n, m and u are the observation number, the equation number and the unknown number, respectively.

Start with a linear or nonlinear functions vector $\mathbf{f}_m(\widehat{\mathbf{y}}, \widehat{\mathbf{x}}) = \mathbf{0}$, we can have a linear mathematical model with a weight matrix ($\mathbf{P} = \sigma_0^2\ \boldsymbol{\Sigma}_{\mathbf{y}}{}^{-1}$) of the observations for $m = n$:

$$\boldsymbol{\varepsilon}_n = \mathbf{A}_{n,u}\ \boldsymbol{\delta}_u - \mathbf{l}_n,\ \mathbf{P}_{n,n}, \tag{1}$$

$$\mathbf{A}_{n,u} = \left.\frac{\partial \mathbf{f}(\widehat{\mathbf{y}}, \widehat{\mathbf{x}})}{\partial \widehat{\mathbf{x}}}\right|_{\widehat{\mathbf{y}}, \widehat{\mathbf{x}} = \mathbf{y}, \mathbf{x}} \quad \text{and} \quad \mathbf{l}_n = \mathbf{f}(\mathbf{y}, \mathbf{x}).$$

Mathematical model between data and unknowns can be established by Taylor expansion for any model. However, if m pieces function vector $\mathbf{f}_m(\widehat{\mathbf{y}}, \widehat{\mathbf{x}}) = \mathbf{0}$ is not transformed into $\widehat{\mathbf{y}}_n - \mathbf{f}_n(\widehat{\mathbf{x}}) = \mathbf{0}$ (for $m = n$), the error in variable solution as in total least squares (TLS) method can be preferred. Therefore, $\mathbf{f}_m(\widehat{\mathbf{y}}, \widehat{\mathbf{x}}) = \mathbf{0}$ (for $m \neq n$) should be differenced as following:

$$\mathbf{B}_{m,n}\ \boldsymbol{\varepsilon}_n - \mathbf{A}_{m,u}\ \boldsymbol{\delta}_u + \mathbf{l}_m = \mathbf{0} \qquad \mathbf{P}_{n,n} \tag{2}$$

where

$$\mathbf{B}_{m,n} = \left. \frac{\partial \mathbf{f}(\widehat{\mathbf{y}}, \widehat{\mathbf{x}})}{\partial \widehat{\mathbf{y}}} \right|_{\hat{\mathbf{y}}, \hat{\mathbf{x}} = \mathbf{y}, \mathbf{x}_0}.$$

Most of science and engineering problems can be modeled as $\widehat{\mathbf{y}} - \mathbf{f}(\widehat{\mathbf{x}}) = \mathbf{0}$ $(m = n)$. Therefore, the functional model named as indirect adjustment method in the adjustment literature [3–7] in geomatics engineering has been preferred in the chapter. The weight matrix $(\mathbf{P}_{n,n})$ of observations (stochastic variables) would be accepted as a unit matrix $\mathbf{P}_{n,n} = \mathbf{I}_{n,n}$ in here for simplicity.

2.1. Objective functions

A generalization for objective functions is $L_p - Norm$ $(p = 1, 2, 3, 4 \ldots, \infty)$ [9, 10]. The first-degree objective function is L_1*-norm estimation* which is accepted as a robust estimation method in just linear models [9–11].

$$\mathbf{i}^T |\boldsymbol{\varepsilon}| \mapsto \min \qquad L_1 - \text{norm } \textit{estimation (Least absolute residuals)}, \tag{3}$$

$$\boldsymbol{\varepsilon}^T \boldsymbol{\varepsilon} \mapsto \min \qquad L_2 - \text{norm } \textit{estimation (Least squares)}, \tag{4}$$

$$|\boldsymbol{\varepsilon}|_{\max} \mapsto \min \qquad L_\infty - \text{norm } \textit{estimation (Minmax absolute residuals)}, \tag{5}$$

$$\mathbf{i} = [1\ 1\ \ldots\ 1]^T.$$

The second-degree objective function is L_2*-norm estimation* which is known as least squares (LS) method and widely used in hard and soft computations.

The last-degree objective function is L_∞*-norm estimation* which is known as minmax method. In fact, the soft computing techniques use this objective while it applies the trial-and-error method in their learning stages. Eq. (1) under L_1*-norm* and L_∞*-norm is* also solved by means of linear programming methods, for this reason; the methods may give several solutions (as being in trial-and-error method) to any interested problem [10, 11].

2.2. Rank deficiencies in linear models

While a rank is a number that indicates a linear independent column, the number of the coefficient matrix of unknowns in a linear equation system, a rank deficiency represents a linear dependent column number (if it is smaller than the row number) of the coefficient matrix. Inconsistency in the solution stage of a linear equation system results from the (rank) deficiencies.

Defining the rank of $\mathbf{A}_{n,u}$ by rank($\mathbf{A}$) $= r$, a condition $r \leq u \leq \min(m, n)$ is always satisfied. In general, $n = m$ in well-known (or the indirect) LS used in many scientific problems.

Denoting the rank defect d letter, we can define two type defects [12].

$$d_s = n - r, \qquad \textit{Surjectivity ("onto" mapping)} \tag{6a}$$

$$d_i = u - r. \qquad \textit{Injectivity ("one - to - one" mapping)} \tag{6b}$$

Objective functions are used to remove the surjectivity defect d_s occurred by the redundant observations. The injectivity defect d_i can consist of three reasons in the estimation problem [12].

Datum defects (d-defects) are closely related to the origin of the spatial system. The defect arises if the data do not carry any information to cover the absolute spatial position of the problem given.

Configuration (Design) defects (c-defects) occur from weak geometric relation among data and unknowns. To avoid the defect, we can be careful and planned when picking data (whose interval or/and place) and choosing the consistent mathematical model (can use auxiliary variables instead of original ones).

Ill-conditioning defects (i-defects) arise from the large intervals among the elements of the coefficient matrix of unknowns. Norming the matrix can reduce ill-conditioning defects but cannot remove it fully. *I-defects* and *c-defects* cannot be separated from each other easily [12].

The defects lead to the failure of any given problem to be solved properly. Since the unknown coefficient matrix cannot be inverted by *regular (ordinary) inverse* methods, we should use *pseudo inverse* to overcome the effects of the defects [8, 9, 13–15]. Eigenvalue and singular value decompositions can be used effectively for the pseudoinverse. Denoting a positive definite symmetric matrix $\mathbf{N}$ (that is always satisfied for $\mathbf{N} = \mathbf{A}^T\mathbf{A}$ or $\mathbf{N} = \mathbf{A}\ \mathbf{A}^T$), its pseudoinverse is:

$$\mathbf{Q}_{u,u} = \mathbf{N}^+ = \mathbf{S}\ \mathbf{\Lambda}^+\ \mathbf{S}^T = \mathbf{V}\ \mathbf{\Sigma}^+\mathbf{U}^T, \qquad \textit{Pseudoinverse of } \mathbf{N} \tag{7a}$$

$$\mathbf{N}_{u,u} = \mathbf{S}\ \mathbf{\Lambda}\ \mathbf{S}^T = \mathbf{U}\ \mathbf{\Sigma}\ \mathbf{V}^T, \qquad \textit{For a positive definite symmetric matrix} \tag{7b}$$

$$\mathbf{\Lambda}^+ = \mathbf{\Sigma}^+ = \begin{bmatrix} \mathbf{\Lambda}_r^{-1} & \mathbf{0}_{r,d} \\ \mathbf{0}_{r,d} & \mathbf{0}_{d,d} \end{bmatrix}.$$

Since $\mathbf{N}_{u,u}$ is a positive definite symmetric matrix in the LS, $\mathbf{S} = \mathbf{U} = \mathbf{V}$. If there is no defect in a matrix $\mathbf{N}$, $\mathbf{N}^{-1} = \mathbf{N}^+$. Therefore, we can use pseudoinverse safely in any given problem [8, 9, 13–15].

2.3. Hard computing

Linearizing from nonlinear functions to their linear form by means of Taylor expansion, a linear equation system is to be handled as Eq. (1). To avoid complicated proofs in the solution of an equation system, the simplified mathematical model can be written in the following (statically rotation invariant [1]) numerical computation form.

$$\mathbf{A}_{n,u}\ \boldsymbol{\delta}_u = \mathbf{l}_n, \qquad \mathbf{P} = \mathbf{I}. \tag{8}$$

Meet two states to solve Eq. (8), $n \le u$ and $n > u$. The solution for the former state $n \le u$ is achieved by means of auxiliary variables vector $\boldsymbol{\lambda}_n$ which can be defined as $\boldsymbol{\delta}_u = \mathbf{A}^T_{u,n}\ \boldsymbol{\lambda}_n$. In fact, the auxiliary vector $\boldsymbol{\lambda}_n$ is named as a Lagrange multipliers vector or an eigenvalues vector in a homogenous equations system in which $\mathbf{l} = \mathbf{0}$ for Eq. (8) [9]. Putting back $\boldsymbol{\delta} = \mathbf{A}^T\ \boldsymbol{\lambda}$ into Eq. (8), we compute $\boldsymbol{\lambda}$ first:

$$\widehat{\boldsymbol{\lambda}}_n = \mathbf{Q}_{n,n}\mathbf{l}_n, \qquad \mathbf{Q}_{n,n} = \left(\mathbf{A}\ \mathbf{A}^T\right)^+. \tag{9}$$

And then $\boldsymbol{\delta}$ and its variance–covariance matrix if we know the statistical uncertainty of observations ($\boldsymbol{\Sigma}_\mathbf{l} = \boldsymbol{\Sigma}_\mathbf{y}$) are calculated by Eq. (10) and the low error propagation, respectively. We can only calculate the variance–covariance matrix of estimations as in Eq. (10) due to $\widehat{\sigma}_0^2 = 0$ and taking $\boldsymbol{\Sigma}_\mathbf{y} = \mathbf{I}$ in the chapter.

$$\widehat{\boldsymbol{\delta}} = \mathbf{A}^T\ \widehat{\boldsymbol{\lambda}} = \mathbf{A}^T\mathbf{Q}\ \mathbf{l}, \qquad \boldsymbol{\Sigma}_{\widehat{\boldsymbol{\delta}}} = \sigma_0^2\ \mathbf{A}^T\mathbf{Q}\,\mathbf{Q}\ \mathbf{A}, \tag{10a}$$

$$\widehat{\mathbf{x}} = \mathbf{x} + \widehat{\boldsymbol{\delta}}, \qquad \boldsymbol{\Sigma}_{\widehat{\mathbf{x}}} = \boldsymbol{\Sigma}_{\widehat{\boldsymbol{\delta}}}, \tag{10b}$$

$$\widehat{\mathbf{x}}^T\widehat{\mathbf{x}} \quad \mapsto \quad \text{min}. \tag{10c}$$

In the state ($n \le u$), $\mathbf{A}\ \widehat{\boldsymbol{\delta}} - \mathbf{l} = \widehat{\boldsymbol{\varepsilon}} = \mathbf{0}$ should be provided. If not, continue solution until $\max\left(|\widehat{\boldsymbol{\delta}}|\right) <= thres = 5e - 12$ (or $\max(|\widehat{\boldsymbol{\varepsilon}}|) <= thres = 5e - 12$) by taking $\mathbf{x} = \widehat{\mathbf{x}}$ in every iteration step. $\widehat{\mathbf{x}}^T\widehat{\mathbf{x}}$ will be the smallest at end of the solution.

Solution to the second state $n > u$ is a situation encountered in many scientific and engineering problems. Multiplying both sides of Eq. (8) by $\mathbf{A}^T_{u,n}$ the normal equation system is established and solved with Eq. (11):

$$\widehat{\boldsymbol{\delta}}_u = \mathbf{Q}\ \mathbf{A}^T\mathbf{l}, \qquad \mathbf{Q}_{u,u} = \left(\mathbf{A}^T\mathbf{A}\ \right)^+, \tag{11a}$$

$$\widehat{\mathbf{x}} = \mathbf{x} + \widehat{\boldsymbol{\delta}}, \qquad \boldsymbol{\Sigma}_{\widehat{\mathbf{x}}} = \widehat{\sigma}_0^2\mathbf{Q}, \tag{11b}$$

$$\widehat{\sigma}_0^2 = \frac{\widehat{\boldsymbol{\varepsilon}}^T\widehat{\boldsymbol{\varepsilon}}}{n - r}, \qquad \textit{A posteriori variance}\ (r = \text{rank}(\mathbf{A})) \tag{11c}$$

$$\widehat{\boldsymbol{\varepsilon}} = \mathbf{A}\ \widehat{\boldsymbol{\delta}} - \mathbf{l}, \tag{11d}$$

$$\widehat{\boldsymbol{\varepsilon}}^T\widehat{\boldsymbol{\varepsilon}} \quad \mapsto \quad \text{min}. \qquad L_2 - \textit{norm estimation (Least Square)} \tag{11e}$$

End the solution if the condition ensured is $\max\left(|\widehat{\boldsymbol{\delta}}|\right) <= thres = 5e - 12$; otherwise, continue the iteration with $\mathbf{x} = \widehat{\mathbf{x}}$.

Relationships between nonlinearity and LS in a multidimensional surface have been shown by Teunissen et al. [1, 2]. The authors argued the relation on some simple examples and gave some analytical solutions for them. But, they highlighted that those types of analytical solutions have not been given for every problem and emphasized that suitable Taylor expansions have been useful to the solution not being transformed into the analytical ones.

3. Geometry of a combination of polynomial and trigonometric functions

These type functions can be used in defining the orbits of artificial satellites (and celestial bodies). Also, the numerical example part of this chapter, to estimate those type functions, will be inspected and applied on a real example. To foresee a model for any problem we should interpret the model parameter and comprehend the geometry of the model (**Figure 1**).

With respect to independent variable time t, a combination function of $p = 1$ degree polynomial and order $q = 1$ trigonometric function(s) [a combination of polynomial degree and trigonometric order (CPT)] to be estimated in the chapter is:

$$\phi_j = a_\phi + b_\phi \ t_j + c_\phi \ \sin\left(d_\phi + e_\phi \ t_j\right), \quad (12)$$

$$\phi_j \in \{X_j, Y_j, Z_j, S_j\}, \ j \in \{1, 2, \ldots, n\}.$$

where $\left(t_j, \phi_j\right)$ are data given. In Eq. (12), translation a_ϕ and slope b_ϕ are elements of a line equation which is a first-order polynomial of CPT function. The other model parameters in the trigonometric part of Eq. (12) are defined as an amplitude c_ϕ, and an initial phase d_ϕ and a frequency (or angular velocity) $e_\phi = 2\pi/T_\phi$ (a period T_ϕ) of a wave (**Figure 1**).

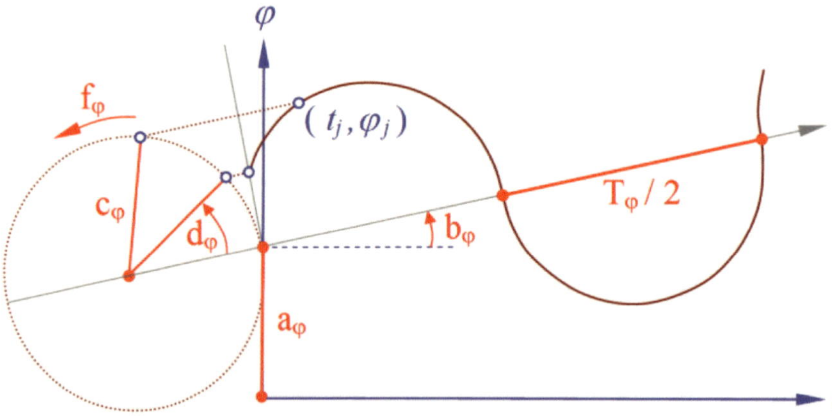

Figure 1. The geometry of a first-degree and first-order combination of polynomial and trigonometric (CPT) function.

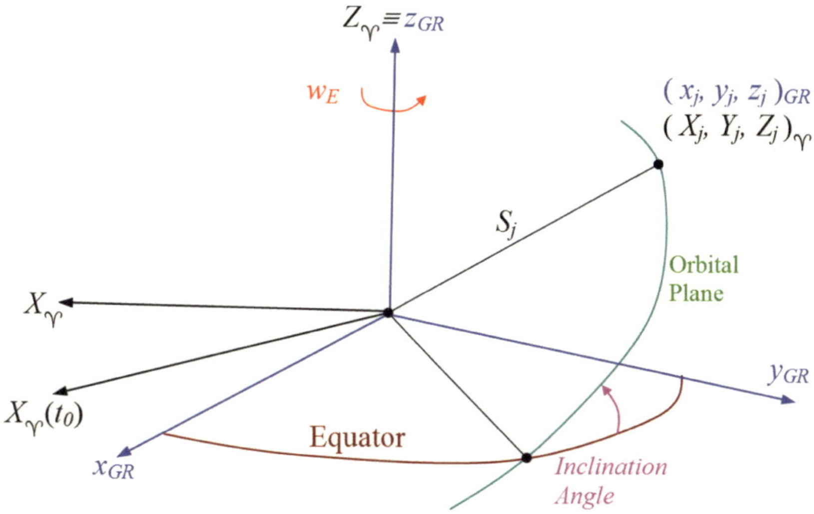

Figure 2. Earth (GR) and space-fixed (ϒ) coordinates for an artificial satellite.

In this chapter, the functions ϕ_j are the coordinate components $\{X_j, Y_j, Z_j\}$ incoming from a precise orbit file and the geometric distances $S_j = \sqrt{X_j^2 + Z_j^2 + Z_j^2}$ as a function of the components. However, nonperiodic earth-fixed coordinates (GR) in the SP3 file should be transformed to the periodic space-fixed coordinates (ϒ); why is Eq. (12) is suitable for the space-fixed coordinates, not earth-fixed ones (as seen from **Figure 3** in the numerical example part) (**Figure 2**)?

For this propose, an easy transformation into any epoch (e.g., it can be taken as the first epoch t_0 of the data) is carried out by:

$$\mathbf{X}_{\gamma,j} = \mathbf{R}_3(\theta_j)\ \mathbf{x}_{GR,j}, \qquad \theta_j = -w_E\ t_j, \tag{13a}$$

$$\mathbf{x}_{GR,j} = \mathbf{R}_3(-\theta_j)\ \mathbf{X}_{\gamma,j}, \qquad \mathbf{R}_3(-\theta_j) = \mathbf{R}_3^T(\theta_j) = \mathbf{R}_3^{-1}(\theta_j), \tag{13b}$$

$$\mathbf{R}_3(\theta_j) = \begin{bmatrix} \cos\theta_j & \sin\theta_j & 0 \\ -\sin\theta_j & \cos\theta_j & 0 \\ 0 & 0 & 1 \end{bmatrix}, \mathbf{X}_{\gamma,j} = \begin{bmatrix} X_j \\ Y_j \\ Z_j \end{bmatrix}_\gamma, \mathbf{x}_{GR,j} = \begin{bmatrix} x_j \\ y_j \\ z_j \end{bmatrix}_{GR}.$$

where w_E and $\mathbf{R}_3$ are in order of the angular velocity of earth and well-known orthogonal rotation matrix around the third axis (**Figure 2**).

A solution of nonlinear Eq. (12) is realized in the following order. Linearizing Eq. (12) by Taylor expansion and omitting the terms greater than or equal to quadratic ones, the linear equation system as given by Eq. (8) is obtained. The explicit form of the Eq. (8) with respect to the approximate values of unknowns for a CPT is:

$$\mathbf{A}_j = \left[1 \quad t_j \quad \sin\left(d_0 + f_0\ t_j\right) \quad c_0 \cos\left(d_0 + e_0\ t_j\right) \quad t_j c_0 \cos\left(d_0 + e_0\ t_j\right)\right], \tag{14a}$$

$$\mathbf{l}_j = \left[\phi_j - \left\{a_0 + b_0\ t_j + c_0\ \sin\left(d_0 + e_0\ t_j\right)\right\}\right],\ j \in \{\ 1, 2, \dots, n\ \}. \tag{14b}$$

We can use a recursive solution for Eq. (14) instead of the batch solution as Eq. (11) because of its solution velocity.

$$\widehat{\boldsymbol{\delta}} = \mathbf{Q}\left(\sum_{j=1}^{n} \mathbf{A}_j^T \mathbf{l}_j\right), \qquad \mathbf{Q} = \left(\sum_{j=1}^{n} \mathbf{A}_j^T \mathbf{A}_j\right)^{-1} = \left(\sum_{j=1}^{n} \mathbf{A}_j^T \mathbf{A}_j\right)^{+}. \tag{15}$$

Continuation of the solution of Eq. (15) can be performed according to Eq. (11). The model given by Eq. (12) is a simple model to determine the satellite orbit motions. For more complicated models, the readers can utilize [19–25] resources.

4. Numerical example

For a nonlinear estimation of CPT functions, some numerical examples are chosen from GNSS {Global navigation satellite systems = GPS (USA) + GLONASS (RU), GALILEO (EU), COMPASS (CHN)} artificial satellite orbits whose coordinates are downloaded from the internet address ftp://ftp.glonass-iac.ru/MCC/PRODUCTS/17091/final/Sta19426.sp3 [26].

For this purpose, two estimation software have been developed in 64Bit Python (in accordance with the 2.7 and 3.6 version) and 32Bit C++ (Code::Blocks) environments to see the convergence rate of the mathematical model given in Eq. (12) [27, 28]. The computed elements of CPT functions for the selected four satellites R01 (GLONASS), G03 (GPS), E01 (GALILEO) and C06 (COMPASS) are summarized in **Table 1** in which they are ordered from the nearest satellite to the farthest one.

The motions of the CPT functions estimated satellites (in **Table 1**) with respect to earth- (left column of **Figure 3**) and space-fixed (right column of **Figure 3**) coordinate systems are demonstrated in **Figure 3**. Moreover, coherence between the estimated CPT function (black solid line) of the C06 satellite and its data points given (colorful circles) is represented in detail in **Figure 4**.

We know that accuracy of precise SP3 file coordinates is about $\sigma_0 = \pm 5\,\text{cm}$. If we compare the value with its estimations given in **Table 1**, we can say that our predicted model is not meet our demands. We should expand the model by raising the degree of polynomial part or/and order of trigonometric part of CPT functions. In fact, we can readily see that the projected model with Eq. (12) will never cover the data. The model is only chosen for this chapter. The more suitable model established on Keplerian orbital elements can be found in the orbit determination literature and in [18–20].

Comparing the solution velocities (from the iteration numbers with respect to $5e-12$ threshold in **Table 1**) in different platforms, we can say that the solution velocities in 64Bit Python are generally better then 32Bit ones.

Sat.	φ	X	Y	Z	Unit	S
	iter	5	4	4	–	10
	iter++	5	4	5	–	9
R01	a_φ	−11.860	2.420	2.771	km	**25,508.091**
(RU)	b_φ	0.028	0.003	0.052	km/h	0.001
	c_φ	25,474.503	11,178.453	22,965.361	km	8.127
	d_φ	−60°54′04.37″	−23°03′56.93″	−30°31′00.32″	deg	42°41′35.87″
	e_φ	31°57′53.92″	−31°57′44.92″	−31°57′56.31″	deg/h	−31°58′13.55″
	T_φ	$11^h05'44.37''$	$11^h15'47.54''$	$11^h15'43.53''$	h	$11^h15'37.46''$
	$\pm\hat{\sigma}_0$	2.902	1.266	2.653	km	0.211
	iter	6	4	4	–	30
	iter++	6	4	4	–	34
G03	a_φ	−17.545	7.332	−1.824	km	**26,561.324**
(USA)	b_φ	−0.132	−0.045	−0.085	km/h	−0.003
	c_φ	24,800.602	17,959.166	21,757.547	km	14.211
	d_φ	66°34′39.73″	−48°06′43.68″	−7°51′47.16″	deg.	−72°24′19.52″
	e_φ	30°04′56.03″	30°04′52.50″	30°05′00.02″	deg./h	−30°08′11.17″
	T_φ	$11^h58'01.91''$	$11^h58'3.31''$	$11^h58'00.32''$	h	$11^h56'44.42''$
	$\pm\hat{\sigma}_0$	4.898	3.648	3.961	km	0.183
	iter	4	5	4	–	32
	iter++	5	5	4	–	93
E01	a_φ	1.866	1.358	−4.673	km	**29,600.332**
(EU)	b_φ	0.052	−0.048	0.004	km/h	−0.000
	c_φ	21,349.011	26,031.918	24,878.432	km	3.720
	d_φ	−35°22′15.46″	85°57′22.67″	−16°23′29.47″	deg.	8°29′58.23″
	e_φ	−25°34′13.06″	−25°34′14.87″	25°34′18.57″	deg./h	−25°42′10.68″
	T_φ	$14^h04'43.81''$	$14^h04'42.81''$	$14^h04'40.78''$	h	$14^h00'22.19''$
	$\pm\hat{\sigma}_0$	1.796	1.428	1.058	km	0.236
	iter	7	6	6	–	28
	iter++	10	7	7	–	29
C06	a_φ	−407.495	228.037	266.178	km	**42,175.353**
(CHN)	b_φ	61.945	−6.376	−7.892	km/h	−0.358
	c_φ	41,716.806	24,832.649	34,235.695	km	227.599
	d_φ	−24°21′36.29″	59°24′50.73″	67°36′9.42″	deg.	18°32′27.85″
	e_φ	−15°07′02.75″	−15°02′16.41″	−15°02′36.74″	deg./h	−14°57′17.11″
	T_φ	$23^h48'48.86''$	$23^h56'22.30''$	$23^h55'49.94''$	h	$24^h04'21.41''$
	$\pm\hat{\sigma}_0$	56.552	38.018	51.901	km	0.491

Table 1. Computed elements of the CPT functions for G03, R01, E01, C06 satellites by $Iter_{MAX}$ = 1000 and *thres* = 5e-12 in loops {iteration numbers of 64Bit Python and 32Bit C++ software in windows are denoted as *iter* and as *iter++* respectively}.

If we chose the threshold as 51e-13, we can see the distinctions of solution convergences between 64Bit and 32Bit running on the estimations of C06 satellite from 1000 (X), 8 (Y), 7 (Z), 1000 (S) in 32Bit C++ and 10 (X), 8 (Y), 6 (Z), 28 (S) in 64Bit Python in Windows. In here, 1000 is the maximum iteration number. If the mathematical model would be more complicated and its data number would be bigger than the number used in the example part, we would see the state more prominently.

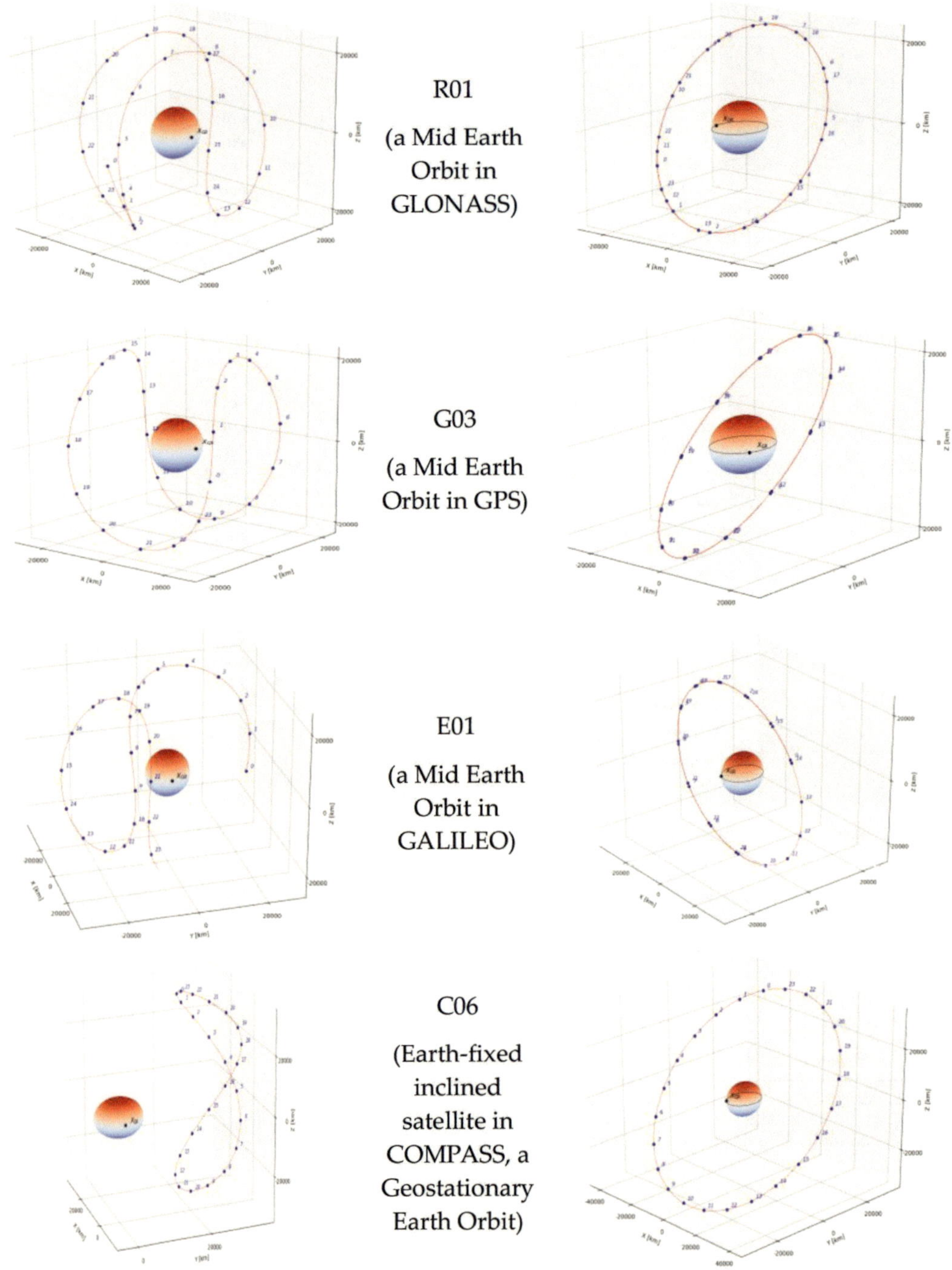

Figure 3. Earth (left) and space (right) fixed orbital traces (see the appendix) with time tags of R01, G03, E02, C06 satellites and the motion of X-coordinate axis shown as GR ($X_{GR}(t_0)$) position on the intersection Greenwich meridian and equator) symbol at t_0 (=2017 April 01 00:00:00).

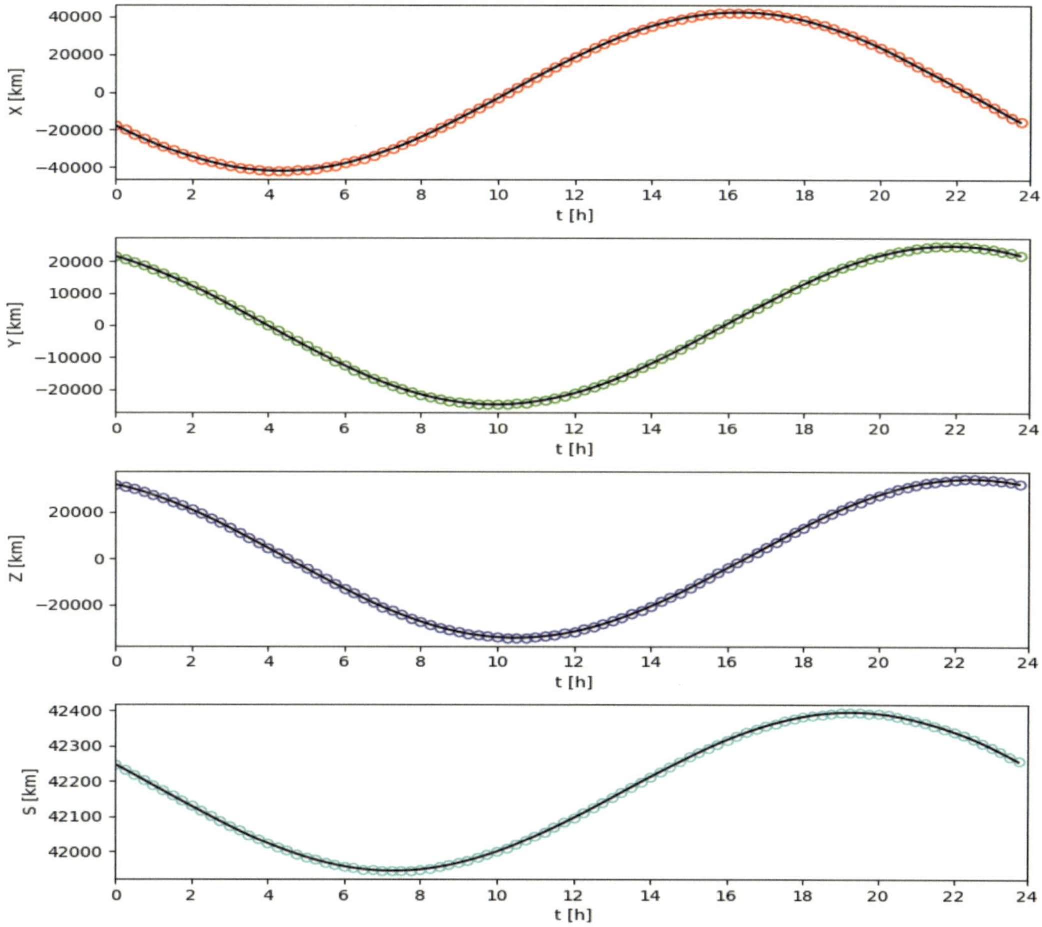

Figure 4. Temporal changing of space fixed coordinates of C06. The circles and solid lines represent the data points and estimated functions under LS respectively for X (red), Y (green), Z (blue), S (cyan).

Approximated values and the loop element for the unknowns are computed following the order in all solutions of the satellites when running in 32Bit C++ and 64Bit Python platforms for the estimations in **Table 1**.

$$a_0 = 0.0, b_0 = 0.0, c_0 = \max\left\{\phi_j\right\}, \quad j \in \{\ 1, 2, \ldots, n = 96\}$$

$$d_0 = \arcsin(\phi_1/c_0), e_0 = \left\{\arcsin(\phi_2/c_0) - \arcsin(\phi_1/c_0)\right\}/\{t_2 - t_1\}$$

$$\mathbf{x} = \begin{bmatrix} a_0 & b_0 & c_0 & d_0 & e_0 \end{bmatrix}^T$$

Maximum iteration number and threshold loop elements are $iter_{MAX} = 1000$ and $thres = 5e - 12$ to break the iteration loop.

Taking the approximated values as $\mathbf{x} = \begin{bmatrix} 0 & 0 & thres & 0 & 0 \end{bmatrix}^T \approx \mathbf{0}^T$ and the same loop elements given above, the iteration numbers are handled as 138 (X), 178 (Y), 78 (Z), 16 (S) in 64Bit Python. For $\mathbf{x} = \begin{bmatrix} 0 & 0 & c_0 & 0 & 0 \end{bmatrix}^T$ they are 33 (X), 235 (Y), 96 (Z), 11 (S) in 64Bit Python. Different approximated value selections cause different iteration numbers (namely convergence rate).

4.1. An expanded model example by an auxiliary cosine wave

Since the estimated standard deviations $\widehat{\sigma}_0 = \{\pm 56.552, \pm 38.018, \pm 51.901, \pm 0.491\}$ (for X, Y, Z, S in **Table 1**) of the CPT functions for the coordinates of the C06 satellite are not statically equal to their expected values ($\sigma_0 = \pm 5\,\text{cm}$), the CPT model should be expanded. As an example, three more unknowns are added to the model given in Eq. (12)

$$\phi_j = a_\phi + b_\phi\ t_j + c_\phi\ \sin\left(d_\phi + e_\phi\ t_j\right) + f_\phi\ \cos\left(g_\phi + h_\phi\ t_j\right)$$

The added terms represent an amplitude f_ϕ, an initial phase g_ϕ and a frequency h_ϕ of a new wave carried by first (sine) wave. After the first estimation with respect to Eq. (12),

we can choose the approximate values of the new parameters as

$$f_0 = \max(abs(\widehat{\boldsymbol{\varepsilon}}))$$

$$g_0 = \arcsin(\widehat{\varepsilon}_1/f_0)$$

$$h_0 = \{\arcsin(\widehat{\varepsilon}_2/f_0) - \arcsin(\widehat{\varepsilon}_1/f_0)\}/\{t_2 - t_1\}$$

from $\widehat{\varepsilon}_j = \left[\phi_j - \left\{\widehat{a} + \widehat{b}\ t_j + \widehat{c}\ \sin\left(\widehat{d} + \widehat{e}\ t_j\right)\right\}\right]$, $j \in \{1, 2, \ldots, n = 96\}$. The approximate value vector of the expanded model by a new wave is:

$$\mathbf{x} = \begin{bmatrix} a_0 & b_0 & c_0 & d_0 & e_0 & f_0 & g_0 & h_0 \end{bmatrix}^T = \begin{bmatrix} \widehat{a} & \widehat{b} & \widehat{c} & \widehat{d} & \widehat{e} & f_0 & g_0 & h_0 \end{bmatrix}^T$$

The approximate values are substituted in the following linearized model as initial values for the loop in LS estimation.

$$\mathbf{l}_j = \left[\phi_j - \left\{a_0 + b_0\ t_j + c_0\ \sin\left(d_0 + f_0\ t_j\right) + e_0 \cos\left(h_0 + g_0\ t_j\right)\right\}\right]$$

$$\mathbf{A}_j^T = \begin{bmatrix} 1 \\ t_j \\ \sin\left(d_0 + e_0\ t_j\right) \\ c_0 \cos\left(d_0 + e_0\ t_j\right) \\ t_j c_0 \cos\left(d_0 + e_0\ t_j\right) \\ \cos\left(g_0 + h_0\ t_j\right) \\ -f_0 \sin\left(g_0 + h_0\ t_j\right) \\ -t_j\ f_0 \sin\left(g_0 + h_0\ t_j\right) \end{bmatrix}$$

After the evaluation, the improved solution for the C06 satellite is represented in **Table 2**. We can readily see the improvements upon the downs of the standard deviations from $\widehat{\sigma}_0 = \{\pm 56.552, \pm 38.018, \pm 51.901, \pm 0.491\}$ (**Table 1**) into $\widehat{\sigma}_0 = \{\pm 0.178, \pm 0.137, \pm 0.191, \pm 0.003\}$

Sat.	φ	X	Y	Z	Unit	S
	iter	27	158	53	–	19
C06	a_φ	223.514	149.296	182.012	km	42,169.956
(CHN)	b_φ	1.515	0.224	0.060	km/h	0.055
	c_φ	42,064.675	66.183	91.527	km	225.161
	d_φ	−25°07′54.78″	11°15′24.23″	3°15′58.76″	deg	19°47′37.54″
	e_φ	−15°02′24.75″	30°04′30.44″	30°04′02.09″	deg/h	−15°02′43.26″
	f_φ	110.710	24,808.034	34,217.994	km	0.838
	g_φ	5°08′57.82″	30°28′55.61″	−22°21′14.60″	deg.	8°56′45.50″
	h_φ	30°10′58.43″	15°02′17.24″	−15°02′19.27″	deg./h	−29°40′58.52″
	$\pm\widehat{\sigma}_0$	0.178	0.137	0.191	km	0.003

Table 2. The results by the expanded model for the C06 satellite in Python 3.6 in windows.

(**Table 2**) in kilometers for the C06 satellite. We can develop the last model more by means of the same manner if we want.

As another example, the expanded model has been trained on the coordinate components of R01 GLONASS satellite by means of Python 3.6 software on Windows. We can also see the improvements upon the downs of the standard deviations (*its iteration numbers*) from $\widehat{\sigma}_0$ = {±2.902 (*5*), ±1.266 (*4*), ±2.653 (*4*), ±0.211 (*10*)} (**Table 1**) to $\widehat{\sigma}_0$ = {±0.226 (*76*), ±0.448 (*32*), ±0.201 (*43*), ±0.037 (*57*)} in kilometers for the R01 satellite.

Condition numbers computed from a rate of maximum and minimum eigenvalues or a rate of singular values under LS are very effective tools for determining the consistency as well. Therefore, a larger condition number can cause larger iteration number (related to convergence rate), We can see those states from the CPT estimations of the C06 satellite with the iteration (*iter*) and condition numbers (*cond*). These are given for X, Y, Z, S as *iter* = *{7, 6, 6,* ***28****}* and *cond* = *{3.4e + 13, 2.0e + 12, 2.8e + 12,* ***1.3e + 09****}* (**Table 1**), and as *iter* = *{27, 158, 53,* ***19****}* and *cond* = *{9.5e + 14, 2.3e + 13, 3.0e + 13,* ***2.6e + 10****}* (**Table 2**).

5. Conclusions

In this chapter, the least squares (LS) estimations of the artificial satellite orbital movements by a combination of polynomial and trigonometric (CPT) functions have been given after a general overview has been made on the hard and soft computations. In practice, the orbital motions are modeled on Keplerian orbital elements. In contrary to this, the coordinate components have been selected for this chapter due to the nonlinear relations of the components and the unknowns which are the elements of CPT functions. The relations cause inconsistencies in the LS solutions. The inconsistencies result from the two injectivity defects, *c-defects* and *i-defects*. We can readily see the defects from the differences of the convergence rates (in other words the iteration numbers) in different computer platforms and architectures as shown in the chapter. The defects are not fully removed as long as not change the mathematic models. However, we can surpass the effects of those defects in part by means of the pseudoinverse based on the eigendecomposition or the singular value decomposition (SVD) as in here. The surjectivity defect (d_s) of the CPT functions not including the datum defects (*d-defects*) was eliminated by the LS objective function.

For the sake of simplicity for readers, a simple CPT function has been chosen at first. After the initial estimation of the function, the estimated errors vector has been found. We have seen that the errors have had a periodic characteristic in time. So, a new wave defining the error characteristic and been able to carry by the first wave has been planned for expanding the CPTs. It is shown that we can expand a CPT function until ensuring statically equivalency between a priory and a posteriori variances. For instance, one may secure the equivalency of the variances if one would expand more by a new wave in the last estimated model in the same manner.

The convergence rates (upon the iteration numbers) of the LS estimation have been inspected according to the threshold (*thres* = *5e-12*) which is a good value for the estimation of the nonlinear CPT function. An algorithm compiled by different compilers and run in different architectures (with 32 Bit or 64 Bit) changes the convergence rate of the estimations in such as the inconsistent scientific problems. It is also observed that the iteration numbers change when the 64-bit Python software is run on Linux platform which has a different framework than Windows. But, the numbers have not been given in the example part of the chapter. Contrary to inconsistency model, namely in a consistent one, the iteration numbers can take equivalence values in all circumstances. Another way to determine the inconsistency of a model is to obtain its condition number which is computed from a rate of maximum and minimum eigenvalues or of singular values under LS. If the condition number is close to one, the projected model is accepted as a consistent model.

We can use the Soft Computing Methods (SCM) if not an exact mathematical relationship between the data and unknowns. The mathematical model is established by the trial-and-error method in training part of SCM by means of arbitrary weights and activation functions depending on SCM expert forecasts. For the solution of the SCM model during the training, we can use least absolute residuals (LAR) and minmax absolute residuals (MAR) objective functions by the linear programming or the LS estimation as in hard computing method (HCM). In the state, the inconsistency problem can erase whatever the solution method (LAR, MAR or LS) is. The inconstancy can be removed by means of experiences gained from HCMs.

Prior information is very important to select a suitable mathematical model for a scientific problem. For example, comparing a priori variance with a posteriori variance at the end of the estimation is a useful warning to the user to determine the correct mathematical model as seen from the expanded model in the example section of the chapter.

In numerical computation, there are two main phenomena which are the mathematical model (as a combination of functional and stochastic models) and objective function. The solution strategy is of no importance if the same mathematical model and objective function are preferred in the same problem of hard computing. All solution strategies always give same results, only their solution time spans can be distinct from each other (**Table 3**).

Acknowledgements

I wish to thank *TÜRKSAT A.S* (https://www.turksat.com.tr/en) supporting this study.

Appendix

Epoch(j)	$t_j - t_0$ [h]	X_j [km]	Y_j [km]	Z_j [km]	t_j [μsec]
0	0.00	−17531.307506	21541.792054	31834.209680	691.175390
1	0.25	−19992.147900	20678.913485	30924.464057	691.487182
2	0.50	−22367.333395	19727.437365	29882.214843	691.798804
3	0.75	−24646.589754	18691.353300	28711.797065	692.110527
4	1.00	−26820.032098	17575.021170	27418.102912	692.422088
5	1.25	−28878.208475	16383.154007	26006.563346	692.733923
6	1.50	−30812.142166	15120.799311	24483.127129	693.045689
7	1.75	−32613.372313	13793.318783	22854.237453	693.357416
8	2.00	−34273.992670	12406.366506	21126.806228	693.668853
9	2.25	−35786.688260	10965.865676	19308.186097	693.980453
10	2.50	−37144.769753	9477.983947	17406.140275	694.291853
11	2.75	−38342.205344	7949.107471	15428.810302	694.603337
12	3.00	−39373.649964	6385.813755	13384.681845	694.914789
13	3.25	−40234.471624	4794.843425	11282.548651	695.226157
14	3.50	−40920.774721	3183.071022	9131.474825	695.537828
15	3.75	−41429.420160	1557.474971	6940.755568	695.849062
16	4.00	−41758.042128	−74.893170	4719.876568	696.160359
17	4.25	−41905.061383	−1706.940006	2478.472204	696.471627
18	4.50	−41869.694990	−3331.562017	226.282782	696.782836
19	4.75	−41651.962330	−4941.677516	−2026.889008	697.094239
20	5.00	−41252.687387	−6530.258698	−4271.222183	697.405373
21	5.25	−40673.497209	−8090.363611	−6496.921900	697.716659
22	5.50	−39916.816531	−9615.167884	−8694.263981	698.028206
23	5.75	−38985.858544	−11097.996030	−10853.639319	698.339619
24	6.00	−37884.611845	−12532.352178	−12965.597898	698.651055
25	6.25	−36617.823589	−13911.950035	−15020.892223	698.962566
26	6.50	−35190.978917	−15230.741942	−17010.519896	699.273893
27	6.75	−33610.276761	−16482.946839	−18925.765115	699.585363
28	7.00	−31882.602129	−17663.077000	−20758.238873	699.896674
29	7.25	−30015.495009	−18765.963379	−22499.917620	700.208082
30	7.50	−28017.116067	−19786.779434	−24143.180195	700.519378
31	7.75	−25896.209299	−20721.063292	−25680.842806	700.830889
32	8.00	−23662.061860	−21564.738140	−27106.191889	701.142117
33	8.25	−21324.461276	−22314.130731	−28413.014646	701.453707

Epoch(j)	t_j–t_0 [h]	X_j [km]	Y_j [km]	Z_j [km]	t_j [μsec]
34	8.50	−18893.650283	−22965.987908	−29595.627135	701.764972
35	8.75	−16380.279535	−23517.491071	−30648.899728	702.076478
36	9.00	−13795.358449	−23966.268499	−31568.279851	702.387737
37	9.25	−11150.204459	−24310.405495	−32349.811865	702.699185
38	9.50	−8456.390949	−24548.452300	−32990.154024	703.010540
39	9.75	−5725.694161	−24679.429736	−33486.592421	703.321863
40	10.00	−2970.039352	−24702.832610	−33837.051887	703.633117
41	10.25	−201.446497	−24618.630817	−34040.103800	703.944417
42	10.50	2568.024186	−24427.268222	−34094.970801	704.255873
43	10.75	5326.326597	−24129.659289	−34001.528424	704.567076
44	11.00	8061.482740	−23727.183555	−33760.303648	704.878433
45	11.25	10761.636054	−23221.677952	−33372.470443	705.189684
46	11.50	13415.103835	−22615.427078	−32839.842340	705.501070
47	11.75	16010.428511	−21911.151470	−32164.862120	705.812547
48	12.00	18536.427516	−21111.993965	−31350.588697	706.124190
49	12.25	20982.241560	−20221.504257	−30400.681303	706.435990
50	12.50	23337.381077	−19243.621724	−29319.381091	706.747258
51	12.75	25591.770667	−18182.656653	−28111.490271	707.058650
52	13.00	27735.791345	−17043.269964	−26782.348927	707.370038
53	13.25	29760.320445	−15830.451549	−25337.809640	707.681516
54	13.50	31656.769034	−14549.497344	−23784.210083	707.992915
55	13.75	33417.116707	−13205.985271	−22128.343727	708.304435
56	14.00	35033.943650	−11805.750146	−20377.428827	708.615955
57	14.25	36500.459878	−10354.857705	−18539.075847	708.927258
58	14.50	37810.531573	−8859.577863	−16621.253481	709.238773
59	14.75	38958.704447	−7326.357314	−14632.253449	709.549980
60	15.00	39940.224086	−5761.791632	−12580.654219	709.861492
61	15.25	40751.053248	−4172.596961	−10475.283826	710.172943
62	15.50	41387.886081	−2565.581420	−8325.181952	710.484439
63	15.75	41848.159196	−947.616367	−6139.561422	710.796071
64	16.00	42130.059795	674.392412	−3927.769279	711.107495
65	16.25	42232.530676	2293.533436	−1699.247596	711.419275
66	16.50	42155.272132	3902.918094	536.505839	711.730566
67	16.75	41898.740894	5495.708947	2769.976807	712.042086
68	17.00	41464.146141	7065.147654	4991.673762	712.353663
69	17.25	40853.442432	8604.582424	7192.166646	712.665021

Epoch(j)	$t_j - t_0$ [h]	X_j [km]	Y_j [km]	Z_j [km]	t_j [μsec]
70	17.50	40069.319938	10107.494927	9362.125250	712.976412
71	17.75	39115.191840	11567.526548	11492.356997	713.287987
72	18.00	37995.179006	12978.503915	13573.844046	713.599417
73	18.25	36714.092016	14334.463621	15597.779572	713.910828
74	18.50	35277.410608	15629.676058	17555.603122	714.221952
75	18.75	33691.260665	16858.668299	19439.034930	714.533584
76	19.00	31962.388796	18016.245942	21240.109087	714.845281
77	19.25	30098.134631	19097.513875	22951.205467	715.156764
78	19.50	28106.400907	20097.895874	24565.080300	715.468390
79	19.75	25995.621474	21013.152993	26074.895308	715.779824
80	20.00	23774.727299	21839.400671	27474.245306	716.091334
81	20.25	21453.110591	22573.124521	28757.184187	716.402951
82	20.50	19040.587166	23211.194732	29918.249189	716.714457
83	20.75	16547.357150	23750.879046	30952.483389	717.026029
84	21.00	13983.964160	24189.854253	31855.456321	717.337474
85	21.25	11361.253075	24526.216175	32623.282652	717.649172
86	21.50	8690.326544	24758.488074	33252.638847	717.960633
87	21.75	5982.500334	24885.627475	33740.777748	718.272113
88	22.00	3249.257698	24907.031342	34085.541002	718.583548
89	22.25	502.202871	24822.539587	34285.369282	718.895203
90	22.50	−2246.986126	24632.436884	34339.310236	719.206528
91	22.75	−4986.605239	24337.452758	34247.024111	719.518326
92	23.00	−7704.972522	23938.759921	34008.787012	719.830016
93	23.25	−10390.476491	23437.970866	33625.491749	720.141620
94	23.50	−13031.624539	22837.132665	33098.646226	720.452998
95	23.75	−15617.091194	22138.719998	32430.369356	720.764247

Table 3. Space Fixed Coordinates of C06 inclined geostationary earth orbit in COMPASS (which is Chinese Global Positioning Satellite System) are transformed with respect to t_0 from earth fixed coordinates downloaded from ftp://ftp.glonass-iac.ru/MCC/PRODUCTS/17091/final/Sta19426.sp3 [26] {t_0 = 2017.04.01–00:00:00 (Civil Calendar) = 1942–518,400 (GPS week—week seconds)}.

Author details

Orhan Kurt

Address all correspondence to: orhnkrt@gmail.com

Geomatics Engineering Department, Kocaeli University, Kocaeli, Turkey

References

[1] Teunissen PJG. The Geometry of Geodetic Inverse Linear Mapping and Non-linear Adjustment, Publications on Geodesy, New Series. Vol. 8, No. 1. Delft: Netherlands Geodetic Commission; 1985. p. 177

[2] Teunissen PJG, Knickmeyer EH. Nonlinearity and least squares. CISM Journal ACSGC. 1988;**42**(4):321-330

[3] Krakiwsky EJ. A Synthesıs of Recent Advances in the Method of Least Squares, Lecture Notes 42, Department of Geo_desy and Geomatics Engineering. Canada: University of New Brunswick; 1975

[4] Mikhail EM. Observations and Least Squares. IEP Series Civil Engineering. New York: Thomas Y. Crowell Company; 1976. ISBN: 0-7002-2481-5

[5] Cross PA. Advanced least squares applied to positioning-fixing, working paper No. 6, University of East London; 1994. ISSN: 0260–9142

[6] Leick A. GPS Satellite Surveying. 2nd ed. A Wiley-Inter-science Publication, New York; 1995. ISBN-10: 0–471–30626-6

[7] Gibbs B. Advanced Kalman Filtering, Least-Squares and Modeling: A Practical Handbook. John Wiley & Sons Inc.; 2011. ISBN 978–0–470-52970-6 (cloth)

[8] Strang G, Borre K. Linear Algebra, Geodesy and GPS. Wellesley College; 1997. ISBN: 9780961408862, 0961408863

[9] Koch KR. Parameter Estimation and Hypothesis Testing in Linear Models: Second, Updated and Enlarged Edition. Springer-Verlag Berlin/Heidelberg; 1999. ISBN 978–3–642-08461-4

[10] Farebrother RW. L1-Norm and L∞-Norm Estimation, An Introduction to the Least Absolute Residuals, the Minimax Absolute Residual and Related Fitting Procedures. Heidelberg/New York/Dordrecht/London: Sprınger; 2013. DOI: 10.1007/978-3-642-36300-9

[11] Barrodale I, Ve Roberts FDK. An improved algorithm for discrete L1-Norm. SIAM Journal of Numerical Analysis. 1973;**10**(5):839-848

[12] Delikaraoglou D. Estimability analyses of the free networks of differential range observations to GPS satellites. In: Grafarend EW, Sanso F, editors. Optimization and Design Geodetic Network. Springer Verlag; 1985. pp. 196-220, 606. ISBN 3-540-15739-5

[13] Press WH, Teukolsky SA, Vetterling WT, Flannery BP. Numerical Recipes in C: The Art of Scientific Computing, 2nd ed. Cambridge University Press: Cambridge; 2002. ISBN 0-521-43108-5

[14] Kiusalaas J. Numerical Methods in Engineering with Python 3. Cambridge/New York/Melbourne/Madrid/Cape Town/Singapore/Sao Paulo/Delhi/Mexico City: Cambridge University Press; 2013. ISBN 978-1-107-03385-6 Hardback

[15] Kurt O. An integrated solution for reducing ill-conditioning and testing the results in non-linear 3D similarity transformations; 2017. http://www.tandfonline.com/doi/full/10.1080/17415977.2017.1337762

[16] Furuta H, Nakatsu K, Hattori H. Applications of soft computing in engineering problems. In: Gregorio Romero Rey, Luisa Martinez Muneta, editors. Modelling Simulation and Optimization. InTechRijeka, Croatia; 2010. ISBN: 978–953–307-048-3. Available online at http://www.intechopen.com/books/modelling-simulationand-optimization/applications-of-soft-computing-in-engineering-problems

[17] Kurhe AB, Satonkar SS, Khanale PB, Ashok S. Soft Computing and Its Applications. BIOINFO Soft Computing. 2011;**1**(1):05-07 Available online at http://www.bioinfo.in/contents.php?id=304

[18] Jain AK, Mao J, Mohiuddin KM. Artificial Neural. Networks: A Tutorial, Computer. IEEE; 1996. pp. 31-44 0018-9162/96/1996. http://www.cogsci.ucsd.edu/~ajyu/Teaching/Cogs202_sp12/Readings/jain_ann96.pdf

[19] Montenbruck O, Gill E. Satellite Orbits: Models, Methods, and Applications. Berlin Heidelberg: Springer-Verlag; 2001. ISBN 3-540-67280-X

[20] Kaula WM. Theory of Satellite Geodesy, Applications of Satellites to Geodesy. Blaisdell Publishing Company, A Division of GİNN and Company. Library of Congress Catalog Card Number: 65-14571; 1966

[21] Vallado DA. Fundamentals of Astrodynamics and Applications. McGraw-Hill; 1997. ISBN: 0-07-066829-9

[22] Seeber G. Satellite Geodesy, 2nd ed. Berlin/New York: Walter de Gruyter; 2003. ISBN: 9783110175493,3-11-017549-5

[23] Xiaofeng FU, Meiping WU. Optimal design of broadcast ephemeris parameters for a navigation satellite system. GPS Solution. 2012;**16**(4):439-448

[24] Xu G, Xu J. Orbits, 2nd Order Singularity-Free Solutions. 2nd ed. Berlin/Heidelberg: Springer-Verlag; 2013. DOI: 10.1007/978-3-642-32793-3

[25] Capderou M. Handbook of Satellite Orbits. Cham/Heidelberg/New York/Dordrecht/London: Springer; 2014. DOI: 10.1007/978-3-319-03416-4

[26] IAC Precise Ephemeris File. http://ftp.glonass-iac.ru/MCC/PRODUCTS/17091/final/Sta19426.sp3 [Accessed: January 01, 2018]

[27] Code::Blocks Homepage. http://www.codeblocks.org/; 2016 [Accessed: January 28, 2016]

[28] Python. https://www.python.org/ [Accessed: January 04, 2017]

Chapter 5

A Gradient Multiobjective Particle Swarm Optimization

Hong-Gui Han, Lu Zhang and Jun-Fei Qiao

Additional information is available at the end of the chapter

http://dx.doi.org/10.5772/intechopen.76306

Abstract

An adaptive gradient multiobjective particle swarm optimization (AGMOPSO) algorithm, based on a multiobjective gradient (MOG) method, is developed to improve the computation performance. In this AGMOPSO algorithm, the MOG method is devised to update the archive to improve the convergence speed and the local exploitation in the evolutionary process. Attributed to the MOG method, this AGMOPSO algorithm not only has faster convergence speed and higher accuracy but also its solutions have better diversity. Additionally, the convergence is discussed to confirm the prerequisite of any successful application of AGMOPSO. Finally, with regard to the computation performance, the proposed AGMOPSO algorithm is compared with some other multiobjective particle swarm optimization (MOPSO) algorithms and two state-of-the-art multiobjective algorithms. The results demonstrate that the proposed AGMOPSO algorithm can find better spread of solutions and have faster convergence to the true Pareto-optimal front.

Keywords: multiobjective particle swarm optimization, multiobjective problem, multiobjective gradient, convergence

1. Introduction

Most of the engineering and practical applications, such as wastewater treatment processes and aerodynamic design problem, often have a multiobjective nature and require solving several conflicting objectives [1–3]. Handling with these multiobjective optimization problems (MOPs), there are always a set of possible solutions which represent the tradeoffs among the objectives known as Pareto-optimal set [4–5]. Evolutionary multiobjective optimization (EMO) algorithms, which are a class of stochastic optimization approaches based on population characteristic, are widely used to solve the MOPs, because a series of Pareto-optimal solutions can be obtained in a single run [6–9]. The multiobjective optimization algorithms are striving to acquire a Pareto-optimal set with good diversity and convergence. The most typical EMO algorithms include the non-dominated sorting genetic algorithm (NSGA) [10] and NSGA-II [11], the strength Pareto evolutionary algorithm (SPEA) [12] and SPEA2 [13], the Pareto archived evolutionary strategy (PAES) [14], the Pareto envelope-based selection algorithm (PESA) [15] and PESA-II [16].

The notable characteristic of particle swarm optimization (PSO) is the cooperation among all particles of a swarm which is attracted toward the global best (gBest) in the swarm and its own personal best (pBest), so the PSO have a better global searching ability [17–19]. Among these EMO algorithms, owing to the high convergence speed and ease of implementation, multiobjective particle swarm optimization (MOPSO) algorithms have been widely used [20–23]. MOPSO can also be applied to multiple difficult optimization problems such as noisy and dynamic problems. However, apart from the archive maintenance, in MOPSO, two issues still remain to be further addressed. The first one is the update of gBest and pBest, because the absolute best solution cannot be selected by the relationship of the non-dominated solutions. Then, the selection of gBest and pBest results in the different flight directions for a particle, which has an important effect on the convergence and diversity of MOPSO [24].

Zheng et al. introduced a novel MOPSO algorithm, which can improve the diversity of the swarm and improve the performance of the evolving particles over some advanced MOPSO algorithms with a comprehensive learning strategy [25]. The experimental results illustrate that the proposed approach performs better than some existing methods on the real-world fire evacuation dataset. In [26], a multiobjective particle swarm optimization with preference-based sort (MOPSO-PS), in which the user's preference was incorporated into the evolutionary process to determine the relative merits of non-dominated solutions, was developed to choose the suitable gBest and pBest. The results indicate that the user's preference is properly reflected in optimized solutions without any loss of overall solution quality or diversity. Moubayed et al. proposed a MOPSO by incorporating dominance with decomposition (D2MOPSO), which proposes a novel archiving technique that can balance the relationship of the diversity and convergence [27]. The analysis of the comparable experiments demonstrates that the D2MOPSO can handle with a wide range of MOPs efficiently. And some other methods for the update of gBest and pBest can be found in [28–31]. Although many researches have been done, it is still a huge challenge to select the appropriate gBest and pBest with the suitable convergence and diversity [32–33].

The second particular issue of MOPSO is how to own fast convergence speed to the Pareto Front, well known as one of the most typical features of PSO. According to the requirement of the fast convergence for MOPSO, many different strategies have been put forward. In [34], Hu et al. proposed an adaptive parallel cell coordinate system (PCCS) for MOPSO. This PCCS is able to select the gBest solutions and adjust the flight parameters based on the measurements of parallel cell distance, potential and distribution entropy to accelerate the convergence of MOPSO by assessing the evolutionary environment. The comparative results show that the self-adaptive MOPSO is better than the other methods. Li et al. proposed a dynamic MOPSO, in which the number of swarms is adaptively adjusted throughout the search process [35]. The dynamic MOPSO algorithm allocates an appropriate number of swarms to support convergence and diversity criteria. The results show that the performance of the proposed dynamic MOPSO algorithm is competitive in comparison to the selected algorithms on some standard benchmark problems. Daneshyari et al. introduced a cultural framework to design a flight parameter mechanism for updating the personalized flight parameters of the mutated particles in [36]. The results show that this flight parameter mechanism performs efficiently in exploring solutions close to the true Pareto front. In the above MOPSO algorithms, the improved strategies are expected to achieve better performance. However, few works have been done to examine the convergence of these MOPSO algorithms [37].

Motivated by the above review and analysis, in this chapter, an adaptive gradient multiobjective particle swarm optimization (AGMOPSO) algorithm, based on a multiobjective gradient (MOG) method, is put forward. This novel AGMOPSO algorithm has faster convergence in the evolutionary process and higher efficiency to deal with MOPs. The proposed AGMOPSO algorithm contains a major contribution to solve MOPs as follows: A novel MOG method is proposed for updating the archive to improve the convergence speed and the local exploitation in the evolutionary process. Unlike some existing gradient methods for single-objective optimization problems [38–40] and MOPs [41], much less is known about the gradient information of MOPs. One of the key feathers of the MOG strategy is that the utilization of gradient information for MOPs is able to obtain a Pareto set of solutions to approximate the optimal Pareto set. In view of the advantages of the MOG strategy, this AGMOPSO algorithm can obtain a good Pareto set and reach smaller testing error with much faster speed. This characteristic makes this method ideal for MOPs.

2. Problem formulation

2.1. Multiobjective problems

A minimize MOP contains several conflicting objectives which is defined as:

$$\begin{aligned} &\text{minimize } F(\mathbf{x}) = \left(f_1(\mathbf{x}), f_2(\mathbf{x}), \cdots, f_m(\mathbf{x})\right)^{\mathrm{T}}, \\ &\text{subject to } \mathbf{x} \in \Omega, \end{aligned} \tag{1}$$

where m is the number of objectives, $\mathbf{x}$ is the decision variable, $f_i()$ is the ith objective function. A decision variable $\mathbf{y}$ is said to dominate the decision vector $\mathbf{z}$, defined as $\mathbf{y}$ dominates $\mathbf{z}$ or $\mathbf{y} \prec \mathbf{z}$, which is indicated as:

$$\forall i : f_i(\mathbf{y}) \leq f_i(\mathbf{z}) \text{ and } \exists j : f_j(\mathbf{y}) < f_j(\mathbf{z}), \tag{2}$$

where $i = 1,2, \ldots, m$, $j = 1,2, \ldots, m$. When there is no solution that can dominate one solution in MOPs, this solution can be used as the Pareto optimal solution. This Pareto optimal solution comprises the Pareto front.

2.2. Particle swarm optimization

PSO is a stochastic optimization algorithm, in which a swarm contains a certain number of particles that the position of each particle can stand for one solution. The position of a particle which is expressed by a vector:

$$\mathbf{x}_i(t) = [x_{i,1}(t), x_{i,2}(t), \ldots x_{i,D}(t)], \tag{3}$$

where D is the dimensionality of the searching space, $i = 1, 2, \ldots, s$; s is the swarm size. Also each particle has a velocity which is represented as:

$$\mathbf{v}_i(t) = [v_{i,1}(t), v_{i,2}(t), \ldots v_{i,D}(t)]. \tag{4}$$

During the movement, the best previous position of each particle is recorded as $\mathbf{p}_i(t) = [p_{i,1}(t), p_{i,2}(t), \ldots, p_{i,D}(t)]$, and the best position obtained by the swarm is denoted as $\mathbf{g}(t) = [g_1(t), g_2(t), \ldots, g_D(t)]$. Based on $\mathbf{p}_i(t)$ and $\mathbf{g}(t)$, the new velocity of each particle is updated by:

$$v_{i,d}(t+1) = \omega v_{i,d}(t) + c_1 r_1 \left(p_{i,d}(t) - x_{i,d}(t)\right) + c_2 r_2 \left(g_d(t) - x_{i,d}(t)\right), \tag{5}$$

where t denotes the tth iteration during the searching process; $d = 1, 2, \ldots, D$ is the dimension in the searching space; ω is the inertia weight; c_1 and c_2 are the acceleration constants and r_1 and r_2 are the random values uniformly distributed in [0, 1]. Then the updating formula of the new position is expressed as:

$$x_{i,d}(t+1) = x_{i,d}(t) + v_{i,d}(t+1). \tag{6}$$

At the beginning of the searching process, the initial position of each particle is randomly generated. As the searching process goes on, the particle swarm may appear as an uneven distribution phenomenon in the evolutionary space.

3. Multiobjective gradient method

The key points of AGMOPSO, compared to the original MOPSO, are that the MOG method is taken into account. In AGMOPSO, the population with N particles intends to search for a set of non-dominated solutions to be stored in an archive with a predefined maximal size.

In MOPSO, the position of each particle can represent the potential solution for the conflicting objectives. The gBest and pBest can guide the evolutionary direction of the whole particle swarm. The position $\mathbf{x}_i$ and velocity $\mathbf{v}_i$ of the ith particle are the D-dimensional vectors $\mathbf{x}_i(0) \in R_D$, $\mathbf{v}_i(0) \in R_D$. The particle updates the velocity and position by the motion trajectory in Eqs. (5) and (6). The external archive $\mathbf{A}(0)$ is initialized as a null set. Meanwhile, the best previous position $\mathbf{p}_i(t)$ is computed by:

$$\mathbf{p}_i(t) = \begin{cases} \mathbf{p}_i(t-1), & \text{if } \mathbf{x}_i(t) \prec \mathbf{p}_i(t-1), \\ \mathbf{x}_i(t), & \text{otherwise}, \end{cases} \tag{7}$$

where $\mathbf{a}_j(t-1) \prec \succ \mathbf{p}_i(t)$ means $\mathbf{x}(t)$ is not dominated by $\mathbf{p}_i(t-1)$. The process of archive $\mathbf{A}(t)$ is updated based on the previous archive $\mathbf{A}(t-1)$ and the best previous position $\mathbf{p}_i(t)$

$$\mathbf{A}(t) = \begin{cases} \mathbf{A}(t-1) \cup \mathbf{p}_i(t), & \text{if } \mathbf{a}_j(t-1) < > \mathbf{p}_i(t), \\ \overline{\mathbf{A}}(t-1) \cup \mathbf{p}_i(t), & \text{otherwise,} \end{cases} \tag{8}$$

where $\mathbf{A}(t) = [\mathbf{a}_1(t), \mathbf{a}_2(t), \ldots, \mathbf{a}_K(t)]^T$, $\mathbf{a}_j(t) = [a_{1,j}(t), a_{2,j}(t), \ldots, a_{D,j}(t)]$, $\overline{\mathbf{A}}(t-1)$ is updated archive which removed the solutions dominated by the best previous position $\mathbf{p}_i(t)$, K is the dimensionality of archive $\mathbf{A}(t)$ which will be changed in the learning process, $\mathbf{a}_j(t-1) < > \mathbf{p}_i(t)$ means $\mathbf{a}_j(t\text{-}1)$ is not dominated by $\mathbf{p}_i(t)$ and $\mathbf{p}_i(t)$ is not dominated by $\mathbf{a}_j(t\text{-}1)$. Moreover, $\mathbf{g}(t)$ is found according to [24].

In AGMOPSO, to enhance the local exploitation, the archive $\mathbf{A}(t)$ is further updated by the MOG method using the gradient information to obtain a Pareto set of solutions that approximates the optimal Pareto set. Without loss of generality, assuming all of the objective functions are differentiable, the directional derivative in $f_i(\mathbf{a}_j(t))$ in a direction $\bar{\mathbf{u}}_j(t)$ at point $\mathbf{a}_j(t)$ is denoted as

$$\nabla_{\bar{\mathbf{u}}_j(t)} f_i(\mathbf{a}_j(t)) = \lim_{\delta \to 0} \left\{ \frac{f_i(\mathbf{a}_j(t) + \delta \bar{\mathbf{u}}_j(t)) - f_i(\mathbf{a}_j(t))}{\delta} \right\}, \tag{9}$$

where $\delta > 0$, $\bar{\mathbf{u}}_j(t) = [\bar{u}_{1,j}(t), \bar{u}_{2,j}(t), \ldots, \bar{u}_{D,j}(t)]$, $i = 1, 2, \ldots, m$; $j = 1, 2, \ldots, K$, and the directional derivative can be rewritten:

$$\nabla_{\bar{\mathbf{u}}_j(t)} f_i(\mathbf{a}_j(t)) = \nabla f_i(\mathbf{a}_j(t)) \bar{\mathbf{u}}_j(t), \tag{10}$$

then, the gradient direction of MOP can be represented as:

$$\nabla_{\bar{\mathbf{u}}_j(t)} \boldsymbol{F}(\mathbf{a}_j(t)) = \left[\nabla_{\bar{\mathbf{u}}_j(t)} f_1(\mathbf{a}_j(t)), \nabla_{\bar{\mathbf{u}}_j(t)} f_2(\mathbf{a}_j(t)), \ldots, \nabla_{\bar{\mathbf{u}}_j(t)} f_m(\mathbf{a}_j(t)) \right]^T, \tag{11}$$

According to Eq. (11), the minimum direction of MOP is calculated as

$$\hat{\mathbf{u}}_i(t) = \frac{\nabla f_i(\mathbf{a}_j(t))}{\|\nabla f_i(\mathbf{a}_j(t))\|},$$
$$\nabla f_i(\mathbf{a}_j(t)) = [\partial f_i(\mathbf{a}_j(t))/\partial a_{1,j}(t), \partial f_i(\mathbf{a}_j(t))/\partial a_{2,j}(t), \ldots, \partial f_i(\mathbf{a}_j(t))/\partial a_{D,j}(t)], \tag{12}$$

and $\|\hat{\mathbf{u}}_i(t)\| = 1$. In addition, the smooth criteria $f_i(\mathbf{a}_j(t))$ are said to be Pareto-stationary at the point $\mathbf{a}_j(t)$ if

$$\sum_{i=1}^{m} \alpha_i(t) \hat{\mathbf{u}}_i(t) = 0, \quad \sum_{i=1}^{m} \alpha_i(t) = 1, \quad \alpha_i(t) \geq 0, \quad (\forall i). \tag{13}$$

The weight vector can be set as

$$\boldsymbol{\alpha}(t) = \frac{1}{\left\| \hat{\mathbf{U}}^T \hat{\mathbf{U}} \right\|^2} \left[\|\hat{\mathbf{u}}_1\|^2, \|\hat{\mathbf{u}}_2\|^2, \ldots, \|\hat{\mathbf{u}}_m\|^2 \right]^T, \tag{14}$$

where $\hat{\mathbf{U}}(t) = [\hat{\mathbf{u}}_1(t), \hat{\mathbf{u}}_2(t), \ldots, \hat{\mathbf{u}}_m(t)]$, $\alpha_i(t) = \|\hat{\mathbf{u}}_i\|^2 / \left\| \hat{\mathbf{U}}^T \hat{\mathbf{U}} \right\|^2$, and $\|\boldsymbol{\alpha}\| = 1$.

To find the set of Pareto-optimal solutions of MOPs, the multi-gradient descent direction is given as follows:

$$\nabla F(\mathbf{a}_j(t)) = \sum_{i=1}^{m} \alpha_i(t)\widehat{\mathbf{u}}_i(t), \quad \sum_{i=1}^{m} \alpha_i(t) = 1, \ \alpha_i(t) \geq 0, \ (\forall i). \tag{15}$$

This multi-gradient descent direction is utilized to evaluate the full set of unit directions. And the archive $\mathbf{A}(t)$ is updated as follows:

$$\bar{\mathbf{a}}_j(t) = \mathbf{a}_j(t) + h \cdot \nabla F(\mathbf{a}_j(t)), \tag{16}$$

where, h is the step size, $\mathbf{a}_j(t)$ and $\bar{\mathbf{a}}_j(t)$ are the jth archive variables before and after the MOG algorithm has been used at time t and the fitness values are updated at the same time.

Moreover, the archive $\mathbf{A}(t)$ can store the non-dominated solutions of AGMOPSO. But the number of non-dominated solutions will gradually increase during the search process. Therefore, to improve the diversity of the solutions, a fixed size archive is implemented in AGMOPSO to record the good particles (non-dominated solutions). During each iteration, the new solutions will be compared with the existing solutions in the archive using the dominating relationship. When a new solution cannot be dominated by the existing solutions in the archive, it will be reserved in the archive. On the contrary, the dominated new solutions cannot be accepted in the archive. If the capacity of the archive reaches the limitation, a novel pruning strategy is proposed to delete the redundant non-dominated solutions to maintain uniform distribution among the archive members.

Assuming that there are K points which will be selected from the archive serve. The maximum distance of the line segment between the first and the end points (namely whole Euclidean distance D_{max}) are obtained. Then, the average distance of the remained K-2 points are set

$$d = D_{max}/(K - 1), \tag{17}$$

where d is the average distance of all points. The average values of d are used to guide to select the non-dominated solutions of more uniform distribution. In addition, for the three objectives, all of the solutions (except the first and the end) are projected to the D_{max}. The points can be reserved, the projective points and the average distance points can be found. However, most projective distances of the adjacent points are not equal to the average distance. Thus, the next point is likely to be selected when it has the distance more closely to the average distance. Once the search process is terminated, the solutions in archive will become the final Pareto front. Taking DTLZ2 as an example, **Figure 1** shows this strategy with three objectives in details.

Local search is a heuristic method to improve PSO performance. It repeatedly tries to improve the current solution by replacing it with a neighborhood solution. In the proposed MOG algorithm, the set of unit directions is described by the normalized combination of the unit directions that map to the intersection points as Eq. (12). Then, each single run of the algorithm can yield a set of Pareto solutions. Experiments demonstrate that the improvements make AGMOPSO effective.

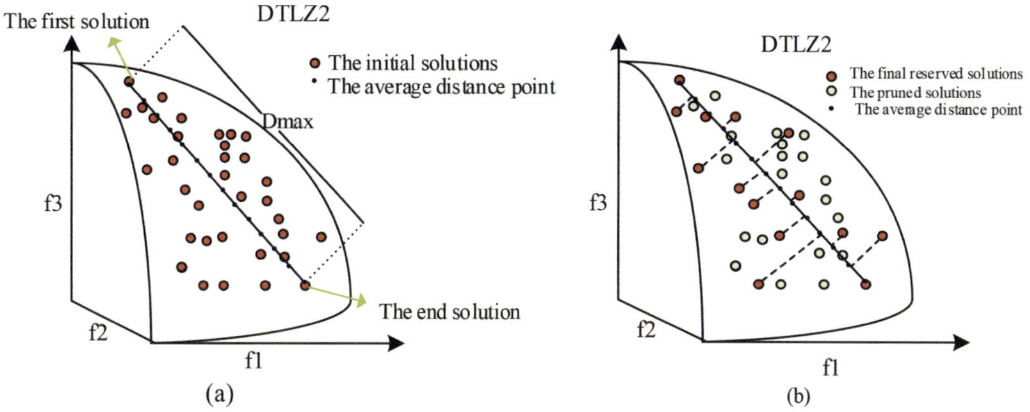

Figure 1. Illustration of points selection procedure. (a) Is the original points and (b) is the selection result of the proposed strategy.

Initializing the flight parameters, population size, the particles positions **x**(0) and velocity **v**(0)	
Loop	
Calculating the fitness value	
Getting the non-dominated solutions	% Eq. (8)
Storing the non-dominated solutions in archive **A**(t)	
Updating the archive using MOG method	% Eq. (16)
If (the number of archive solutions exceed capacity)	
Pruning the archive	
End	
Selecting the gBest from the archive **A**(t)	
Calculating the flight parameters	
Updating the velocity $\mathbf{x}_i$(t) and position $\mathbf{v}_i$(t)	% Eqs. (5–6)
End loop	

Table 1. AMOPSO algorithm.

In MOPSO, it is desired that an algorithm maintains good spread of solutions in the non-dominated solutions as well as the convergence to the Pareto-optimal set. In this AGMOPSO algorithm, an estimate of density is designed to evaluate the density of solutions surrounding it. It calculates the overall Euclidean distance values of the solutions, and then the average distance of the solutions along each of the objectives corresponding to each objective is calculated. This method is able to get a good spread result under some situations to improve the searching ability. And the pseudocode of AGMOPSO is presented in **Table 1**.

4. Simulation results and analysis

In this section, three ZDT and two DTLZ benchmark functions are employed to test the proposed of AGMOPSO. This section compares the proposed AGMOPSO with four state-of-the-art MOPSO algorithms—adaptive gradient MOPSO (AMOPSO) [41], crowded distance MOPSO (cdMOPSO) [32], pdMOPSO [31] and NSGA-II [11].

4.1. Performance metrics

To demonstrate the performance of the proposed AGMOPSO algorithm, two different quantitative performance metrics are employed in the experimental study.

1. Inverted generational distance (*IGD*):

$$IGD(\boldsymbol{F}^*, \boldsymbol{F}) = \sum_{\boldsymbol{x} \in \boldsymbol{F}^*} \boldsymbol{mindis}(\boldsymbol{x}, \boldsymbol{F}) / |\boldsymbol{F}^*|, \tag{18}$$

where $mindis(\boldsymbol{x}, \boldsymbol{F})$ is the minimum Euclidean distance between the solution $\boldsymbol{x}$ and the solutions in $\boldsymbol{F}$. A smaller value of $IGD(\boldsymbol{F}^*, \boldsymbol{F})$ demonstrates a better convergence and diversity to the Pareto-optimal front.

2. Spacing (*SP*):

$$SP = \sqrt{\frac{1}{K-1} \sum_{i=1}^{K} (\bar{d} - d_i)}, \tag{19}$$

where d_i is the minimum Euclidean distance between ith solution and other solutions, K is the number of non-dominated solutions, $\bar{d}$ is the average distance of the all Euclidean distance d_i.

4.2. Parameter settings

All the algorithms have three common parameters: the population size N, the maximum number of non-dominated solutions K and iterations T. Here, $N = 100$, $K = 100$ and $T = 3000$.

4.3. Experimental results

The experimental performance comparisons of the cdMOPSO algorithm on ZDTs and DTLZs are shown in **Figures 2–6**. Seen from **Figures 2–6**, the non-dominated solutions obtained by the proposed AGMOPSO algorithm can approach to the Pareto Front appropriately and maintain

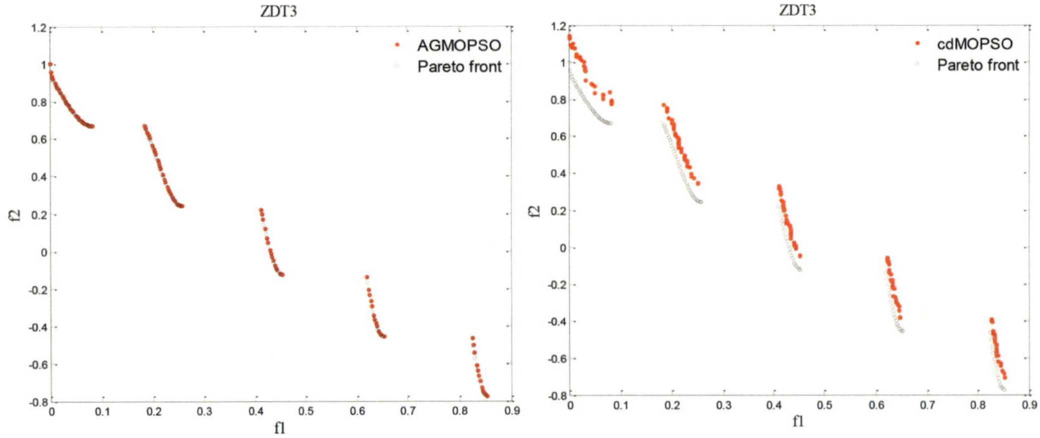

Figure 2. The Pareto front with non-dominated solutions obtained by the two multiobjective algorithms for ZDT3 function.

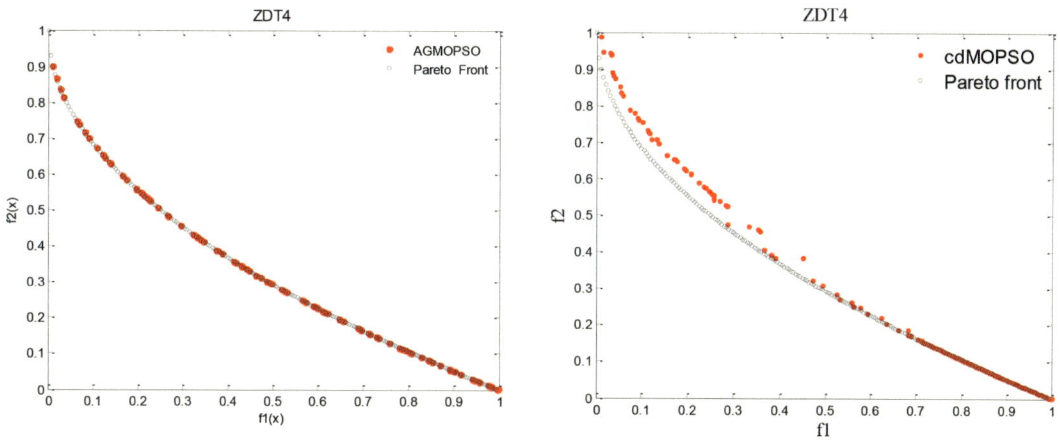

Figure 3. The Pareto front with non-dominated solutions obtained by the two multiobjective algorithms for ZDT4 function.

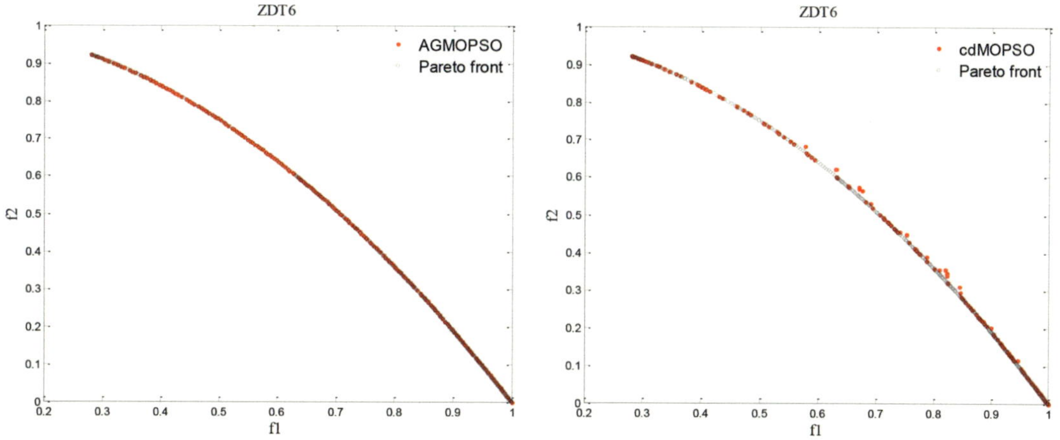

Figure 4. The Pareto front with non-dominated solutions obtained by the two multiobjective algorithms for ZDT6 function.

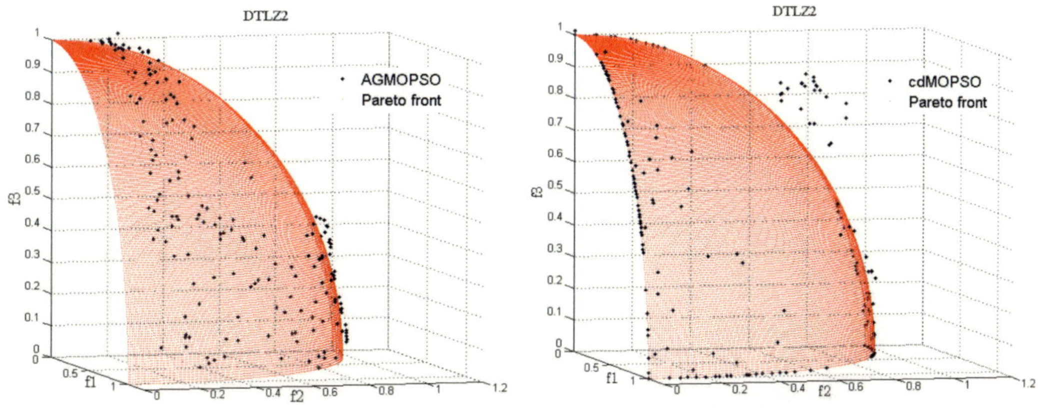

Figure 5. The Pareto front with non-dominated solutions obtained by the two multiobjective algorithms for DTLZ2 function.

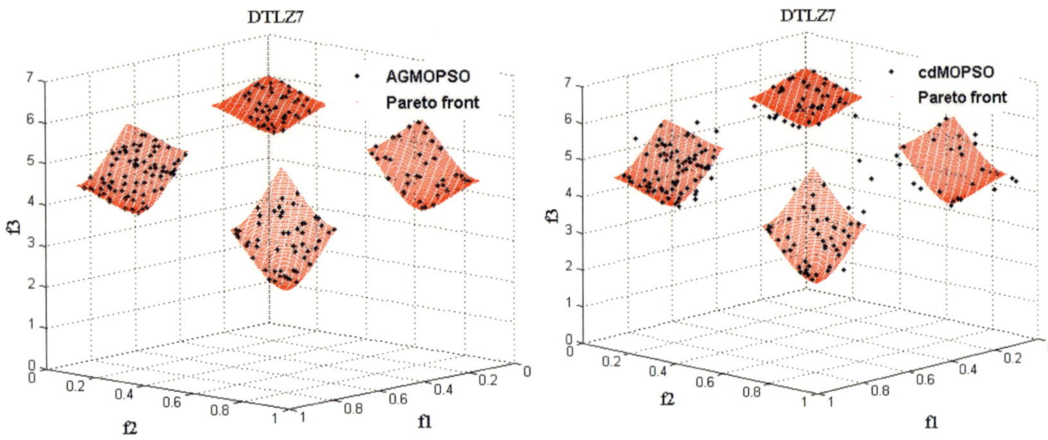

Figure 6. The Pareto front with non-dominated solutions obtained by the two multiobjective algorithms for DTLZ7 function.

Function	Index	AGMOPSO	AMOPSO	pdMOPSO	cdMOPSO	NSGA-II
ZDT3	Best	0.00149	0.00425	0.2019	0.003109	0.005447
	Worst	0.00697	0.00832	0.4265	0.028986	0.006105
	Mean	0.00433	0.00632	0.3052	0.003063	0.005834
	std	0.00297	0.00527	0.1003	0.007131	0.000202
ZDT4	Best	3.0194	2.7133	3.3980	4.9760	0.00462
	Worst	5.1522	5.0543	4.9760	6.3610	0.11166
	Mean	3.7933	3.8943	4.0330	5.9120	0.016547
	std	1.5133	2.7401	1.6510	4.5180	0.031741
ZDT6	Best	0.2046	0.0936	2.2310	0.000897	0.01119
	Worst	0.7834	0.9154	2.8790	0.003627	0.01498
	Mean	0.4878	0.5433	2.4690	0.002988	0.01286
	std	0.0242	0.0236	0.8169	0.0001543	0.001004
DTLZ2	Best	0.0477	0.0519	0.1330	0.0322	0.07830
	Worst	0.3913	0.3425	0.3690	0.2067	0.2740
	Mean	0.1058	0.1878	0.2070	0.1015	0.1059
	std	0.0060	0.0132	0.0413	0.0134	0.008383
DTLZ7	Best	0.05766	0.02044	0.00796	0.00701	0.00614
	Worst	0.32803	0.10295	0.07678	0.05439	0.03208
	Mean	0.01985	0.04573	0.04831	0.02856	0.01799
	std	0.00139	0.00312	0.00289	0.00165	0.00129

Table 2. Comparisons of different algorithms for *IGD*.

Function	Index	AGMOPSO	AMOPSO	pdMOPSO	cdMOPSO	NSGA-II
ZDT3	Best	0.023475	0.097811	0.099654	0.10356	0.081569
	Worst	0.087874	0.416626	0.487126	0.87449	0.106568
	Mean	0.067451	0.245931	0.198551	0.59684	0.092216
	std	0.012873	0.050937	0.079442	0.22468	0.008415
ZDT4	Best	0.030914	0.039825	0.069564	0.139577	0.031393
	Worst	0.078011	0.193765	0.233794	0.300951	0.044254
	Mean	0.049923	0.078821	0.186698	0.204573	0.038378
	std	0.001092	0.004517	0.063757	0.095562	0.003837
ZDT6	Best	0.010981	0.008739	0.009935	0.012396	0.006851
	Worst	0.100551	0.088535	0.023766	0.040205	0.010127
	Mean	0.034127	0.040251	0.010683	0.034569	0.008266
	std	0.009756	0.007341	0.003021	0.003884	0.000918
DTLZ2	Best	0.1438	0.0943	0.0569	0.0932	0.021456
	Worst	0.6893	0.8947	0.6991	0.5897	0.7314
	Mean	0.0398	0.4631	0.4721	0.3562	0.4162
	std	0.00764	0.03401	0.02964	0.01772	0.03655
DTLZ7	Best	0.1958	0.1047	0.0932	0.1347	0.0632
	Worst	0.9032	0.9355	0.8361	0.9307	0.7466
	Mean	0.0502	0.0493	0.4459	0.5972	0.4191
	std	0.01097	0.03201	0.00896	0.2133	0.00796

Table 3. Comparisons of different algorithms for *SP*.

a greater diversity than other compared algorithms. Experimental results in **Figures 2–4** show that the proposed AGMOPSO algorithm is superior to the cdMOPSO algorithm in diversity performance and can approach the Pareto Front. In addition, the results in **Figures 5** and **6** show that the proposed AGMOPSO algorithm can obtain a better performance on the three-objective benchmark problems with accurate convergence and the preferable diversity.

In order to show the experimental performance in details, the experimental results, which contain the best, worst, mean and standard deviations of *IGD* and *SP* based on the two-objective of ZDTs and the three-objective of DTLZs are listed in **Tables 2** and **3**, respectively.

Moreover, the experimental results in **Tables 2** and **3** include the details of the four evolutionary algorithms. To illustrate the significance of the findings, the comparing results for the performance index is analyzed as follows:

1. Comparison of *IGD* index: From **Table 2**, the proposed AGMOPSO algorithm is superior to other MOPSO algorithms in terms of the results of *IGD*. Firstly, in the two-objective of ZDTs instances, the AGMOPSO can have better mean deviations of *IGD* than other four evolutionary algorithms on ZDT3 and ZDT4. It is indicated that the MOG method has

played a vital role on the algorithm. Meanwhile, compared with NSGA-II [11], the proposed AGMOPSO has better *IGD* index performance of accuracy and stability for the two-objective of ZDTs (except ZDT4). Second, in the three-objective of DTLZs instances, the AGMOPSO is superior to other four algorithms in terms of the mean deviations value of *IGD*. According to the comparisons between the AGMOPSO and other four evolutionary algorithms, it is demonstrated that the proposed AGMOPSO is the closest to the true front and nearly enclose the entire front, which means the proposed AGMOPSO algorithm achieves the best convergence and divergence.

2. Comparison of *SP* index: The comparison of *SP* among the proposed AGMOPSO algorithm and other compared algorithms was shown in **Table 3**. Firstly, in the two-objective of ZDTs instances, the AGMOPSO can have better mean deviations and best deviations of *SP* than other four evolutionary algorithms ZDT3 and ZDT4. Meanwhile, compared with NSGA-II [11], the proposed AGMOPSO has better *SP* index performance of diversity for the two-objective of ZDTs (except ZDT6). From the results in **Table 3**, the comparison of the *SP* between the proposed AGMOPSO algorithm illustrate that the MOG method can have better effect on the diversity performance than other existing methods. Secondly, in the three-objective of DTLZs instances, the proposed AGMOPSO algorithm has the best *SP* performance on the DTLZ2 and DTLZ7 than the other four compared algorithms. In addition, to verify the effect of the MOG method, the proposed AGMOPSO can obtain a set of non-dominated solutions with greater diversity and convergence than NSGA-II on instances (except ZDT4 and ZDT6). Therefore, the proposed AGMOPSO algorithm can obtain more accurate solutions with better diversity on the most ZDTs and DTLZs.

5. Conclusion

A novel method, named AGMOPSO, is proposed to solve MOPs, which underlies MOG to accelerate the solution convergence and deletes the redundant solutions in the archive by the equidistant partition principle. Meanwhile, the convergence analysis and convergence conditions of AGMOPSO are also carefully investigated for the successful applications. Based on the theoretical analysis and the experimental results, the proposed AGMOPSO algorithm with the local search strategy MOG is a novel method for solving theses MOPs. The comparisons of the different indexes also demonstrate that the proposed AGMOPSO algorithm is superior to the other algorithms for most of ZDTs and DTLZs.

Author details

Hong-Gui Han[1,2]*, Lu Zhang[1,2] and Jun-Fei Qiao[1,2]

*Address all correspondence to: rechardhan@sina.com

1 College of Electronic and Control Engineering, Beijing University of Technology, Beijing, China

2 Beijing Key Laboratory of Computational Intelligence and Intelligent System, Beijing, China

References

[1] Wang Y, Li HX, Yen GG, Song W. MOMMOP: Multiobjective optimization for locating multiple optimal solutions of multimodal optimization problems. IEEE Transactions on Cybernetic's. 2015;**45**(4):830-843

[2] Yuan W, You XG, Xu J, et al. Multiobjective optimization of linear cooperative spectrum sensing: Pareto solutions and refinement. IEEE Transactions on Cybernetics. 2016;**46**(1): 96-108

[3] Brockhoff D, Zitzler E. Objective reduction in evolutionary multiobjective optimization: Theory and applications. Evolutionary Computation. 2009;**17**(2):135-166

[4] Hu XB, Wang M, Paolo ED. Calculating complete and exact Pareto front for multiobjective optimization: A new deterministic approach for discrete problems. IEEE Transactions on Cybernetics. 2013;**43**(3):1088-1101

[5] Martin D, Rosete A, Alcala FJ, Francisco H. A new multiobjective evolutionary algorithm for mining a reduced set of interesting positive and negative quantitative association rules. IEEE Transactions on Evolutionary Computation. 2014;**18**(1):54-69

[6] Jiang SY, Yang SX. An improved multiobjective optimization evolutionary algorithm based on decomposition for complex Pareto fronts. IEEE Transactions on Cybernetics. 2016;**46**(2):421-437

[7] Nag K, Pal NR. A multiobjective genetic programming-based ensemble for simultaneous feature selection and classification. IEEE Transactions on Cybernetics. 2016;**46**(2): 499-510

[8] He D, Wang L, Yu L. Multi-objective nonlinear predictive control of process systems: A dual-mode tracking control approach. Journal of Process Control. 2015;**25**:142-151

[9] Zhang X, Tian Y, Jin Y. A knee point-driven evolutionary algorithm for many-objective optimization. IEEE Transactions on Evolutionary Computation. 2015;**19**(6):761-776

[10] Srinivas N, Deb K. Muiltiobjective optimization using nondominated sorting in genetic algorithms. Evolutionary Computation. 1994;**2**(3):221-248

[11] Deb K, Pratap A, Agarwal S, et al. A fast and elitist multiobjective genetic algorithm: NSGA-II. IEEE Transactions on Evolutionary Computation. 2002;**6**(2):182-197

[12] Zitzler E, Thiele L. An evolutionary algorithm for multiobjective optimization: The strength pareto approach. Tech. Rep. TIK-Report. Zurich, Switzerland: Swiss Federal Institute Technology (ETH); 1998

[13] Zitzler E, Laumanns M, Thiele L. SPEA2: Improving the strength Pareto evolutionary algorithm. Tech. Rep. TIK-Report 103. Zurich, Switzerland: Swiss Federal Institute Technology (ETH); 2001

[14] Knowles JD, Corne DW. Approximating the nondominated front using the Pareto archived evolution strategy. Evolutionary Computation. 2000;**8**(2):149-172

[15] Corne DW, Knowles JD, Oates MJ. The Pareto envelope-based selection algorithm for multiobjective optimization. International Conference on Parallel Problem Solving from Nature. 2000:839-848

[16] Qu BY, Suganthan P, Das S. A distance-based locally informed particle swarm model for multimodal optimization. IEEE Transactions on Evolutionary Computation. 2013;**17**(3): 387-402

[17] Ali H, Shahzad W, Khan FA. Energy-efficient clustering in mobile ad-hoc networks using multi-objective particle swarm optimization. Applied Soft Computing. 2012;**12**(7):1913-1928

[18] AlRashidi MR, El-Hawary ME. A survey of particle swarm optimization applications in electric power systems. IEEE Transactions on Evolutionary Computation. 2009;**13**(4):913-918

[19] Elhossini A, Areibi S, Dony R. Strength pareto particle swarm optimization and hybrid EA-PSO for multi-objective optimization. Evolutionary Computation. 2010;**18**(1):127-156

[20] Zhang Y, Gong D, Zhang J. Robot path planning in uncertain environment using multi-objective particle swarm optimization. Neurocomputing. 2013;**103**:172-185

[21] Li J, Zhang JQ, Jiang CJ, Zhou MC. Composite particle swarm optimizer with historical memory for function optimization. IEEE Transactions on Cybernetics. 2015;**45**(10):2350-2363

[22] Hu WW, Tan Y. Prototype generation using multiobjective particle swarm optimization for nearest neighbor classification. IEEE Transactions on Cybernetics. 2016;**46**(12):2719-2731

[23] Xue B, Zhang MJ, Browne WN. Particle swarm optimization for feature selection in classification: A multi-objective approach. IEEE Transactions on Cybernetics. 2013;**43**(6): 1656-1671

[24] Hu M, Weir JD, Wu T. An augmented multi-objective particle swarm optimizer for building cluster operation decisions. Applied Soft Computing. 2014;**25**:347-359

[25] Zheng YJ, Ling HF, Xue JY, Chen SY. Population classification in fire evacuation: A multiobjective particle swarm optimization approach. IEEE Transactions on Evolutionary Computation. Feb. 2014;**18**(1):70-81

[26] Lee KB, Kim JH. Multiobjective particle swarm optimization with preference-based sort and its application to path following footstep optimization for humanoid robots. IEEE Transactions on Evolutionary Computation. 2013;**17**(6):755-766

[27] Al Moubayed N, Petrovski A, McCall J. D2MOPSO: MOPSO based on decomposition and dominance with archiving using crowding distance in objective and solution spaces. Evolutionary Computation. 2014;**22**(1):47-77

[28] Zhu Q, Lin Q, Chen W, et al. An external archive-guided multiobjective particle swarm optimization algorithm. IEEE Transactions on Cybernetics. To be published. DOI: 10.1109/TCYB. 2017.2710133

[29] Zheng YJ, Chen SY. Cooperative particle swarm optimization for multiobjective transportation planning. Applied Intelligence. 2013;**39**(1):202-216

[30] Agrawal S, Panigrahi BK, Tiwari MK. Multiobjective particle swarm algorithm with fuzzy clustering for electrical power dispatch. IEEE Transactions on Evolutionary Computation. 2008;**12**(5):529-541

[31] Yen GG, Leong WF. Dynamic multiple swarms in multiobjective particle swarm optimization. IEEE Transactions on Systems, Man and Cybernetics, Part A: Systems and Humans. 2009;**39**(4):890-911

[32] Helwig S, Branke J, Mostaghim S. Experimental analysis of bound handling techniques in particle swarm optimization. IEEE Transactions on Evolutionary Computation. 2013; **17**(2):259-271

[33] Zhang Y, Gong DW, Ding Z. A bare-bones multi-objective particle swarm optimization algorithm for environmental/economic dispatch. Information Sciences. 2012;**192**(1):213-227

[34] Hu W, Yen GG. Adaptive multiobjective particle swarm optimization based on parallel cell coordinate system. IEEE Transactions on Evolutionary Computation. 2015;**19**(1):1-18

[35] Li X, Yao X. Cooperatively coevolving particle swarms for large scale optimization. IEEE Transactions on Evolutionary Computation. 2012;**16**(2):210-224

[36] Daneshyari M, Yen GG. Cultural-based multiobjective particle swarm optimization. IEEE Transactions on Systems, Man, and Cybernetics, Part B: Cybernetics. 2011;**41**(2):553-567

[37] Britto A, Pozo A. Using reference points to update the archive of MOPSO algorithms in many-objective optimization. Neurocomputing. 2014;**127**(1):78-87

[38] Chakraborty P, Das S, Roy GG. On convergence of the multi-objective particle swarm optimizers. Information Sciences. 2011;**181**(8):1411-1425

[39] Garcia NJ, Olivera AC, Alba E. Optimal cycle program of traffic lights with particle swarm optimization. IEEE Transactions on Evolutionary Computation. 2013;**17**(6):823-839

[40] Gong M, Cai Q, Chen X. Complex network clustering by multiobjective discrete particle swarm optimization based on decomposition. IEEE Transactions on Evolutionary Computation. 2014;**18**(1):82-97

[41] Mousa AA, El-Shorbagy MA, Abd-El-Wahed WF. Local search based hybrid particle swarm optimization algorithm for multiobjective optimization. Swarm and Evolutionary Computation. 2012;**3**(1):1-14

Chapter 6

Piecewise Parallel Optimal Algorithm

Zheng Hong Zhu and Gefei Shi

Additional information is available at the end of the chapter

http://dx.doi.org/10.5772/intechopen.76625

Abstract

This chapter studies a new optimal algorithm that can be implemented in a piecewise parallel manner onboard spacecraft, where the capacity of onboard computers is limited. The proposed algorithm contains two phases. The predicting phase deals with the open-loop state trajectory optimization with simplified system model and evenly discretized time interval of the state trajectory. The tracking phase concerns the closed-loop optimal tracking control for the optimal reference trajectory with full system model subject to real space perturbations. The finite receding horizon control method is used in the tracking program. The optimal control problems in both programs are solved by a direct collocation method based on the discretized Hermite–Simpson method with coincident nodes. By considering the convergence of system error, the current closed-loop control tracking interval and next open-loop control predicting interval are processed simultaneously. Two cases are simulated with the proposed algorithm to validate the effectiveness of proposed algorithm. The numerical results show that the proposed parallel optimal algorithm is very effective in dealing with the optimal control problems for complex nonlinear dynamic systems in aerospace engineering area.

Keywords: optimal control, parallel onboard optimal algorithm, discretizing Hermite–Simpson method, nonlinear dynamic system, aerospace engineering

1. Introduction

Space tether system is a promising technology over decades. It has wide potential applications in the space debris mitigation & removal, space detection, power delivery, cargo transfer and other newly science & technic missions. Recently, there is continuous interest in the space tether systems, in leading space agencies such as, NASA's US National Aeronautics and Space Administration, ESA's European Space Agency, and JAXA's Japan Aerospace Exploration Agency [1]. Their interest technologies include the electrodynamic tether (EDT) propulsion technology, retrieval of tethered satellite system, multibody tethered system and space elevator system. Compared with existing technologies adopted by large spacecraft such as the rocket or thruster, the space tether technology has the advantages of fuel-efficiency (little or no propellant required), compact size, low mass, and ease-of-use [2]. These advantages make it reasonable to apply the space tethered system for deorbiting the fast-growing low-cost micro/nano-satellites and no-fuel cargo transfer. The difficulty associated with space tether system is to control & suppress its attitudes during a mission process for the technology to be functional and practical. Many works have been devoted to solving this problem, and one effort is to use the optimal control due to its good performances in the complex and unstable nonlinear dynamic systems. In this chapter, a new piecewise onboard parallel optimal control algorithm is proposed to control and suppress the attitudes of the space tether system. To test its validity, two classical space tether systems, the electrodynamic tether system (EDT) and partial space elevator (PSE) system are considered and tested.

An EDT system with constant tether length is underactuated. The electric current is the only control input if there are no other active forces such as propulsion acting on the ends of an EDT. The commonly adopted control strategy in the literature is the current regulation using energy-based feedback in this underactuated control problem. Furthermore, many efforts have been done to solve this problem with optimal control. Stevens and Baker [3] studied the optimal control problem of the EDT libration control and orbital maneuverer efficiency by separating the fast and slow motions using an averaged libration state dynamics as constraints instead of instantaneous dynamic constraints in the optimal control algorithm. The instantaneous states are propagated from the initial conditions using the optimal control law in a piecewise fashion. Williams [4] treated the slow orbital and fast libration motions separately with two different discretization schemes in the optimal control of an EDT orbit transfer. The differential state equations of the libration motion are enforced at densely allocated nodes, while the orbital motion variables are discretized by a quadrature approach at sparsely allocated nodes. The two discretization schemes are unified by a specially designed node mapping method to reflect the coupling nature of orbital and libration motions. The control reference, however, is assumed known in advance.

A PSE system is consisted with one main satellite and two subsatellites (climber & end body) connected to each other by tether(s). The difficulty associated to such a system is to suppress the libration motion of the climber and the end body. This libration is produced by the moving climber due to the Coriolis force, which will lead the system unstable. While the climber is fast moving along the tether, the Coriolis force will lead to the tumbling of the PSE system. Thus, the stability control for suppressing such a system is critical for a successful climber transfer mission. To limit the fuel consuming, tension control is widely used to stable the libration motion of the space tethered system due to it can be realized by consuming electric energy only [5]. Many efforts have been devoted to suppressing the libration motion of space tethered system such as, Wen et al. [6] stabled the libration of the tethered system by an analytical feedback control law that accounts explicitly for the tension constraint. The study shows good computational effect, and the proposed method requires small data storage ability. Ma et al. [7] used adaptive saturated sliding mode control to suppress the attitude angle in the deployment period of the space tethered system. Optimal control [8, 9] is also proved as a way to overcome the libration issue. The above tension control schemes are helpful for both two-body and three-body tethered system. Up to data, limited devotions have been done on the libration suppression of a PSE system using tension control only. Williams used optimal control to design the climber's speed function of a climber for a full space elevator [10]. Modeled by simplified dynamic equations, an optimal control problem is solved, and the solution results in zero in-plane libration motion of the ribbon in the ending phase of climber motion. The study shows that to eliminate the in-plane oscillations by reversing the direction of the elevator is possible. Kojima et al. [11] extended the mission function control method to eliminate the libration motion of a three body tethered system. The proposed method is effective when the total tether length is fixed and the maximum speed of the climber no more than 10 *m/s*. Although these efforts are useful to suppress the libration motion of the PSE system, it still difficult to control the attitudes of such a system in the transfer period.

To overcome the challenges in aforementioned works, we propose a parallel onboard optimal algorithm contains two phases. Phase 1 concerns the reference state trajectory optimization within a given time interval, where an optimal control model is formulated based on the timescale separation concept [3, 12] to simplify the dynamic calculations of the EDT & PSE system. An open-loop optimal state trajectory is then obtained by minimizing a cost function subject to given constraints. The state trajectory of paired state and control input variables is solved approximately by the direct collocation method [13] that is based on the Hermite-Simpson method [14]. In this phase, the simplified dynamic model is by used. Phase 2 concerns the tracking of the open-loop optimal state trajectory within the same interval. A closed-loop optimal control problem is formulated in a quadrature form to track the optimal state trajectory obtained in phase 1. Unlike phase 1, all the major perturbative forces are included, and more realistic geomagnetic and gravitational field models are considered. While the system is running the process in phase 2 with one CPU, the next phase 1 calculation is running in another CPU with data modification based on the errors obtained in the last calculation program. The simulation results demonstrate the effectiveness of the approach in fast satellite deorbit by EDTs in equatorial orbit. Furthermore, for fast transfer period of the partial space elevator, the propose method also shows good effect on suppression the libration angles of the climber and the end body with tension control only.

2. Optimal control algorithm

2.1. Control scheme

Assume two CUPs are used to process the calculation. CUP-1 is used to determine the open-loop optimal control trajectory of dynamic states employing the simple dynamic equations. The obtained optimal state trajectory will be tracked by CPU-2 using closed-loop RHC. While the system is tracking the *i*-th interval, the $(i + 1)$-th optimal trajectory is being calculated in CPU-1. Once the tracking for the *i*-th interval is finished (implemented by CPU-2), the real finial state S_i will be stored in the memory and the $(i + 1)$-th optimal trajectory can be tracked. By repeating the above process, the optimal suppression control problem is solved in a parallel piecewise manner until the transfer period is over.

The calculation of the optimal trajectory in CPU-1 is a prediction state trajectory whose initial state is estimated as $\tilde{S}_i^{i+1} = \tilde{S}_i^i + \frac{1}{2}e_{i-1}$, where $\tilde{S}_i^i$ denotes the estimated finial state of the i-th interval obtained by CUP-1, the superscript and subscript denote the internal number and the states node number. $\tilde{S}_i^{i+1}$ is the estimated initial state of the (i + 1)-th interval, and $e_{i-1} = S_{i-1} - \tilde{S}_{i-1}^{i-1}$ is the error between the real and the estimated final state of the (i-1)-th interval, the data used to calculate the error is picked up from the memory. For the first interval, i = 1, $S_0 = \tilde{S}_0^1$ and $e_0 = 0$. The computational diagram of the entire control strategy is given as shown in **Figure 1**.

2.2. Open-loop control trajectory

The libration angles of the climber and the end body are required to be kept between the desired upper/lower bounds in the climber transfer process. To make the calculation convenient and simple. The accessing process is divided into a series of intervals, such that, the transfer process $[t_i, \ \ t_{i+1}]$ is discretized into n intervals, where t_i and t_{i+1} are the initial and final time, respectively. t_{i+1} can be obtained in terms of the transfer length of the climber and its speed. To make calculation convenient to be realized in practical condition, the transfer process is divided evenly. The optimal trajectory should be found to satisfy the desired cost functions for each time intervals as

$$J_i = \int_{t_i}^{t_{i+1}} \Pi(\boldsymbol{x}, \boldsymbol{u}) \ dt \tag{1}$$

subject to the simplified dynamic equations. All the errors between simple model and the entire model are regarded as perturbations. The above cost function minimization problem is solved by a direct solution method, which uses a discretization scheme to transform the

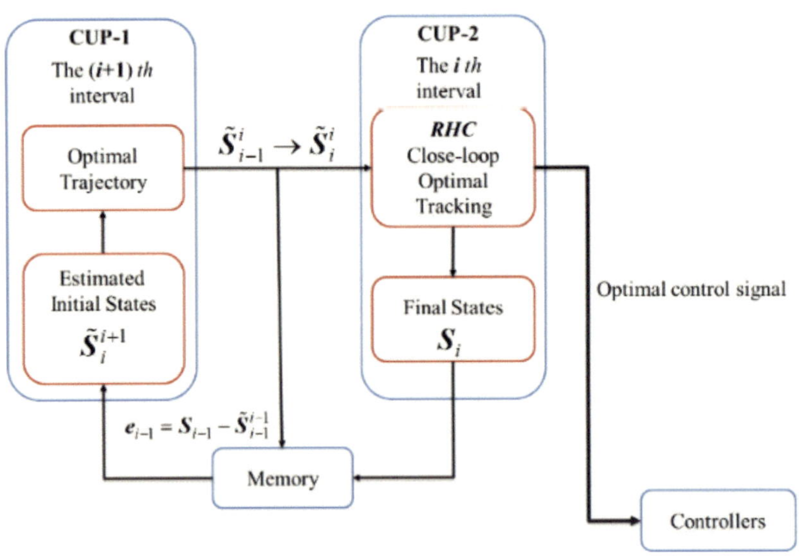

Figure 1. Control scheme.

continuous problem into a discrete parameter optimization problem of nonlinear programming within the interval, to avoid the difficulty usually encountered when standard approaches are used on derivation of the required conditions for optimality [15]. There are a number of efficient discretization schemes, such as, Hermite-Legendre-Gauss-Lobatto method [16] and Chebyshev pseudospectral method [8], in the literature for discretizing the continuous problem. In the current work, a direct collocation method, based on the Hermite-Simpson scheme [14, 17], is adopted because of its simplicity and accuracy.

Assume that the time interval $[t_i, t_{i+1}]$ is discretized into n subintervals with $n + 1$ nodes at the discretized time τ_k $(k = 0, 1, \ldots., n)$.

$$\gamma_k = \tau_{k+1} - \tau_k, \sum_{k=1}^{n} \gamma_k = t_{i+1} - t_i \tag{2}$$

The state vectors and control inputs are discretized at $n + 1$ nodes, $\boldsymbol{x}_0, \boldsymbol{x}_1, \boldsymbol{x}_2, \ldots, \boldsymbol{x}_n$ and $\upsilon_0, \upsilon_1, \upsilon_2, \ldots, \upsilon_n$. Further, denote the state vectors and the control inputs at mid-points between adjacent nodes by $\boldsymbol{x}_{0.5}, \boldsymbol{x}_{1.5}, \boldsymbol{x}_{2.5}, \ldots, \boldsymbol{x}_{n-0.5}$ and $\upsilon_{0.5}, \upsilon_{1.5}, \upsilon_{2.5}, \ldots, \upsilon_{n-0.5}$ and the mid-point state vectors, $\boldsymbol{x}_{k+0.5}$, can be derived by the Hermite interpolation scheme,

$$\boldsymbol{x}_{k+0.5} = \frac{1}{2}(\boldsymbol{x}_k + \boldsymbol{x}_{k+1}) + \frac{\gamma}{8}[\Gamma(\boldsymbol{x}_k, \upsilon_k, \tau_k) - \Gamma(\boldsymbol{x}_{k+1}, \upsilon_{k+1}, \tau_{k+1})] \tag{3}$$

Accordingly, the cost function in Eq. (1) can be discretized by the Simpson integration formula as

$$J \cong \frac{\gamma}{6}\sum_{k=0}^{n-1} [\Pi(\boldsymbol{x}_k, \upsilon_k, \tau_k) + 4\Pi(\boldsymbol{x}_{k+0.5}, \upsilon_{k+0.5}, \tau_{k+0.5}) + \Pi(\boldsymbol{x}_{k+1}, \upsilon_{k+1}, \tau_{k+1})] \tag{4}$$

The nonlinear constraints based on the tether libration dynamics, the first-order states, can also be denoted by discretized equations using the Simpson integration formula, such that

$$\frac{\gamma}{6}[\Pi(\boldsymbol{x}_k, \upsilon_k, t_k) + 4\Pi(\boldsymbol{x}_{k+0.5}, \upsilon_{k+0.5}, \tau_{k+0.5}) + \Pi(\boldsymbol{x}_{k+1}, \upsilon_{k+1}, \tau_{k+1})] + \boldsymbol{x}_k - \boldsymbol{x}_{k+1} = \boldsymbol{0} \tag{5}$$

The left-hand side of Eq. (5) is also named as the Hermite-Simpson Defect vector in the literature. Finally, the discretization process is completed by replacing the constraints for the initial states and the continuous box constraints with the discretized constraints,

$$\boldsymbol{x}_0 = \boldsymbol{x}_{start}, \boldsymbol{x}_{\min} \leq \boldsymbol{x}_k \leq \boldsymbol{x}_{\max}, \upsilon_{\min} \leq \upsilon_k \leq \upsilon_{\max}, \upsilon_{\min} \leq \upsilon_{k+0.5} \leq \upsilon_{\max} \tag{6}$$

The minimization problem of a continuous cost function is now transformed to a nonlinear programming problem. It searches optimal values for the programming variables that minimize the discretized form of cost function shown in Eq. (4) while satisfying the constraints of Eqs. (5) and (6). The subscript index "*k*" will be refreshed in the next time interval.

2.3. Closed-loop optimal control for tracking open-loop optimal state trajectory

The RHC is implemented by converting the continuous optimal control problem into a discrete parameter optimization problem that can be solved analytically. Like the open-loop trajectory optimization problem, the same direct collocation based on the Hermite-Simpson method is used to discretize the RHC problem.

By using the similar notations of discretization above, the cost function is discretized using the Simpson integration formula as

$$G = \frac{1}{2}\delta\boldsymbol{x}_{i+1}^T\boldsymbol{S}\delta\boldsymbol{x}_{i+1}$$
$$+\frac{\gamma}{12}\sum_{k=0}^{n-1}\left[\delta\boldsymbol{x}_k^T\boldsymbol{Q}\delta\boldsymbol{x}_k + 4\delta\boldsymbol{x}_{k+0.5}^T\boldsymbol{Q}\delta\boldsymbol{x}_{k+0.5} + \delta\boldsymbol{x}_{k+1}^T\boldsymbol{Q}\delta\boldsymbol{x}_{k+1} + \boldsymbol{R}\left(\delta\boldsymbol{\upsilon}_k^2 + 4\delta\boldsymbol{\upsilon}_{k+0.5}^2 + \delta\boldsymbol{\upsilon}_{k+1}^2\right)\right] \tag{7}$$

and the constraints are discretized into

$$\delta\boldsymbol{x}_k - \delta\boldsymbol{x}_{k+1} + \frac{\gamma}{6}\left[\boldsymbol{A}_k\delta\boldsymbol{x}_k + \boldsymbol{B}_k\delta\upsilon_k + 4\boldsymbol{A}_{k+0.5}\delta\boldsymbol{x}_{k+0.5} + 4\boldsymbol{B}_{k+0.5}\delta\upsilon_{k+0.5} + \boldsymbol{A}_{k+1}\delta\boldsymbol{x}_{k+1} + \boldsymbol{B}_{k+1}\delta\upsilon_{k+1}\right] = \boldsymbol{0} \tag{8}$$

$$\delta\boldsymbol{x}_k = \boldsymbol{x}(\tau_k) - \boldsymbol{x}_{opt}(\tau_k) \tag{9}$$

$$\delta\boldsymbol{x}_{k+0.5} = \frac{1}{2}(\delta\boldsymbol{x}_k + \delta\boldsymbol{x}_{k+1}) + \frac{\gamma}{8}(\boldsymbol{A}_k\delta\boldsymbol{x}_k + \boldsymbol{B}_k\delta\upsilon_k - \boldsymbol{A}_{k+1}\delta\boldsymbol{x}_{k+1} - \boldsymbol{B}_{k+1}\delta\upsilon_{k+1}) \tag{10}$$

where, $\boldsymbol{A}_k = \boldsymbol{A}(\tau_k)$, $\boldsymbol{A}_{k+0.5} = \boldsymbol{A}(\tau_{k+0.5})$, $\boldsymbol{B}_k = \boldsymbol{B}(\tau_k)$, $\boldsymbol{B}_{k+0.5} = \boldsymbol{B}(\tau_{k+0.5})$.

The derivation of Eq. (8a) finally leads to a quadratic programming problem to find a programming vector $\boldsymbol{Z} = \left[\delta\boldsymbol{x}_0^T\ \delta\boldsymbol{x}_1^T\ \ldots\ \delta\boldsymbol{x}_n^T\ \delta\upsilon_0\ \delta\upsilon_1\ \ldots\ \delta\upsilon_n\ \delta\upsilon_{0.5}\ \delta\upsilon_{1.5}\ \ldots\ \delta\upsilon_{n-0.5}\right]^T$, which minimizes the cost function:

$$G = \frac{1}{2}\boldsymbol{Z}^T\boldsymbol{M}\boldsymbol{Z} \tag{11}$$

subject to.

$$\boldsymbol{C}\boldsymbol{Z} = \boldsymbol{X} \quad \boldsymbol{X} = \left[\delta\boldsymbol{x}_0^T\ 0\ 0\ \ldots\ 0\right]^T \tag{12}$$

where the matrices $\boldsymbol{C}$ and $\boldsymbol{M}$ are given in the Appendix.

It is easy to find the solution analytically to this standard quadratic programming problem by [24].

$$\boldsymbol{Z}^* = \boldsymbol{M}^{-1}\boldsymbol{C}^T\left(\boldsymbol{C}\boldsymbol{M}^{-1}\boldsymbol{C}^T\right)^{-1}\boldsymbol{X} \tag{13}$$

and the control correction at the current time can be obtained as

$$\delta\upsilon(t_i) = \boldsymbol{V}\boldsymbol{Z}^* = \boldsymbol{V}\boldsymbol{M}^{-1}\boldsymbol{C}^T\left(\boldsymbol{C}\boldsymbol{M}^{-1}\boldsymbol{C}^T\right)^{-1}\boldsymbol{X} \triangleq \boldsymbol{K}(t_i, n, t_h)\left[\boldsymbol{x}(t_i) - \boldsymbol{x}_{opt}(t_i)\right] \tag{14}$$

where the row vector $\boldsymbol{V}$ is defined to "choose" the target value from the optimal solution, and the position of "1" in the row vector $\boldsymbol{V}$ is the same as the position of $\delta\upsilon_0$ in the column vector $\boldsymbol{Z}$. Finally, the control input of the closed-loop control, $\upsilon(t_i)$, is

$$v(t_i) = v_{opt}(t_i) + \delta v(t_i) = v_{opt}(t_i) + \boldsymbol{K}(t_i, n, t_h)\left[\boldsymbol{x}(t_i) - \boldsymbol{x}_{opt}(t_i)\right] \tag{15}$$

It is apparent that the closed-loop control law derived here is a linear proportional feedback control law, and the feedback gain matrix $\boldsymbol{K}$ is a function of time. Without any explicit integration of differential equations, $\boldsymbol{K}$ can either be determined offline or online depending on the computation and restoration ability onboard the satellite.

It is worth to point out some advantages of this approach. Firstly, the matrices $\boldsymbol{M}$ and $\boldsymbol{C}$ are both formulated by the influence matrices at certain discretization notes ($\boldsymbol{A}_k$, $\boldsymbol{B}_k$, $\boldsymbol{A}_{k+0.5}$, $\boldsymbol{B}_{k+0.5}$). If t_i and t_i are both set to be coincident with the discretization nodes used in the open-loop control problem, then most of the influence matrices calculated previously can be used directly in the tracking control process to reduce computational efforts. This is the advantage of using the same discretization method in the current two-phased optimal control approach. Secondly, the matrix $\boldsymbol{M}$ is unchanged if treating the terminal horizon time $t_i + t_h$ as the time-to-go and keep the future horizon interval $[t_i, t_{i+1}]$ unchanged. This means the inverse of $\boldsymbol{M}$ could be calculated only once in the same interval. It is attractive for the online implementation of HRC, where the computational effort is critical. As the entire interval can be discretized into small intervals, if these intervals are sufficiently small relative to the computing power of the satellite, then the calculation process can be carried by the onboard computer. Furthermore, for small intervals, the computation for the open-loop optimal trajectory of the next interval can be done by CPU-1 while the tracking is still in process, see **Figure 1**. This makes of the proposed optimal suppression control a parallel online implementation, which is another advantage of this control scheme.

3. Cases study

3.1. Parallel optimal algorithm in attitudes control of EDT system

In order to test the validity of the proposed optimal algorithm, a case study of the attitudes control of EDT system in aerospace engineering is used. The obtained results are compared with some existing control methods.

3.1.1. Problem formulation

The EDT system's orbital motion is generally described in an Earth geocentric inertial frame ($OXYZ$) with the origin O at the Earth's centre, see **Figure 2(a)**. The X-axis directs to the point of vernal equinox, the Z-axis aligns with the Earth's rotational axis, and the Y-axis completes a right-hand coordinate system, respectively. The equation of orbital dynamic motion can be written in the form of Gaussian perturbation [18], which is a set of ordinary differential equations of six independent orbital elements (a, Ω, i, e_x, e_y, φ)

$$\frac{da}{dt} = \frac{2}{n\sqrt{1-e^2}}\left(\sigma_z e \sin \nu + \sigma_x \frac{p}{r}\right) \tag{16}$$

$$\frac{d\Omega}{dt} = \frac{\sigma_y r \sin u}{na^2\sqrt{1-e^2}\sin i} \tag{17}$$

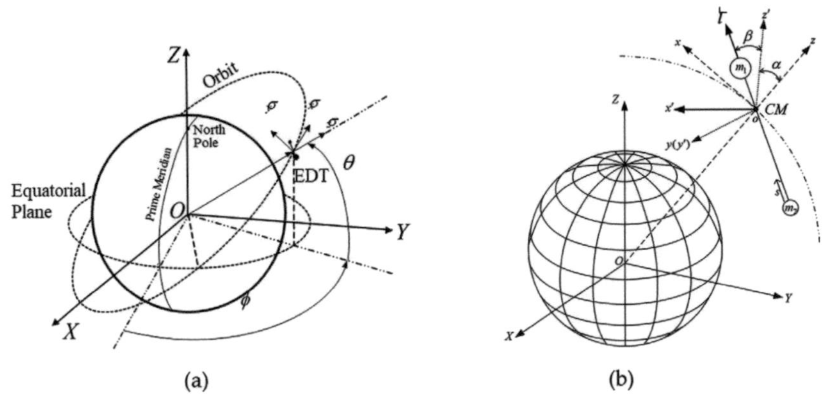

Figure 2. Illustration of coordinate system for the EDT's orbital (a) and libration (b) motion.

$$\frac{di}{dt} = \frac{\sigma_y r \cos u}{na^2\sqrt{1-e^2}} \tag{18}$$

$$\frac{de_x}{dt} = \frac{\sqrt{1-e^2}}{na}\left\{\sigma_z \sin u + \sigma_x\left[\left(1+\frac{r}{p}\right)\cos u + \frac{r}{p}e_x\right]\right\} + \frac{d\Omega}{dt}e_y \cos i \tag{19}$$

$$\frac{de_y}{dt} = \frac{\sqrt{1-e^2}}{na}\left\{-\sigma_z \cos u + \sigma_x\left[\left(1+\frac{r}{p}\right)\sin u + \frac{r}{p}e_y\right]\right\} - \frac{d\Omega}{dt}e_x \cos i \tag{20}$$

$$\frac{d\varphi}{dt} = n - \frac{1}{na}\left[\sigma_z\left(\frac{2r}{a} + \frac{\sqrt{1-e^2}}{1+\sqrt{1-e^2}}e\cos v\right) - \sigma_x\left(1+\frac{r}{p}\right)\frac{\sqrt{1-e^2}}{1+\sqrt{1-e^2}}e\sin v\right] - \frac{\sigma_y r \cos i \sin u}{na^2\sqrt{1-e^2}\sin i} \tag{21}$$

The components of perturbative accelerations are defined in a local frame. The components σ_x and σ_z are in the orbital plane, and σ_z is the radial component pointing outwards. The out-of-plane component σ_y completes a right hand coordinate system. The components of perturbative accelerations depend on the tether attitude and the EDT' orbital dynamics is coupled with the tether libration motion.

The libration motion of a rigid EDT system is described in an orbital coordinate system shown in **Figure 2(b)**. The z-axis of the orbital coordinate system points from the Earth's center to the CM of the EDT system, the x-axis lies in the orbital plane and points to the direction of the EDT orbital motion, perpendicular to the z-axis. The y-axis completes a right-hand coordinate system. The unit vectors along each axis are expressed as $\vec{e}_{ox}$, $\vec{e}_{oy}$, and $\vec{e}_{oz}$, respectively. Then, the instantaneous attitudes of the EDT system are described by an in-plane angle α (pitch angle, rotating about the y-axis) followed by an out-of-plane angle β (roll angle, rotating about the x'-axis, the x-axis after first rotating about the y-axis). Thus, the equations of libration motion of the EDT system can be derived as,

$$\ddot{\alpha} + \ddot{v} - 2(\dot{\alpha} + \dot{v})\dot{\beta}\tan\beta + 3\mu r^{-3}\sin\alpha\cos\alpha = \frac{Q_\alpha}{\tilde{m}L^2\cos^2\beta} \tag{22}$$

$$\ddot{\beta} + (\dot{\alpha} + \dot{\nu})^2 \sin\beta\cos\beta + 3\mu r^{-3}\cos^2\alpha\sin\beta\cos\beta = \frac{Q_\beta}{\tilde{m}L^2} \tag{23}$$

where $\tilde{m} = \left[m_1 m_2 + (m_1 + m_2)m_t/3 + m_t^2/12\right]m_{EDT}^{-1}$ is the equivalent mass, (Q_α, Q_β) are the corresponding perturbative torques by the perturbative forces to be discussed below.

The perturbative accelerations $(\sigma_x, \sigma_y, \sigma_z)$ and torques (Q_α, Q_β) in Eq. (15) are induced by multiple orbital perturbative effects, namely, (i) the electrodynamic force exerting on a current-carrying EDT due to the electromagnetic interaction with the geomagnetic field, (ii) the Earth's atmospheric drag, (iii) the Earth's non-homogeneity and oblateness, (iv) the lunisolar gravitational perturbations, and (v) the solar radiation pressure, respectively. The EDT system is assumed thrust-less during the deorbit process, while the atmosphere, geomagnetic and ambient plasma fields are assumed to rotate with the Earth at the same rate. The geodetic altitude, instead of geocentric altitude, should be used in the evaluation of the environmental parameters, such as, atmospheric and plasma densities, to realistically account for the Earth's ellipsoidal surface, such that,

$$h_g = r - r_{po}\left(1 - e_E^2\cos^2\theta\right)^{-1/2} \tag{24}$$

where the polar radius r_{po} and the Earth's eccentricity e_E are provided by NASA [19].

Moreover, the local strength of geomagnetic field is described by the IGRF2000 model [20–22] in a body-fixed spherical coordinates of the Earth, such that

$$\begin{aligned} B_\phi &= \frac{1}{\sin\theta}\sum_{n=1}^{\infty}\left(\frac{r_0}{r}\right)^{n+2}\sum_{m=0}^{n} m\left[g_n^m \sin(m\phi) - h_n^m\cos(m\phi)\right]P_n^m(\theta_c) \\ B_\theta &= \sum_{n=1}^{\infty}\left(\frac{r_0}{r}\right)^{n+2}\sum_{m=0}^{n}\left[g_n^m\cos(m\phi) + h_n^m\sin(m\phi)\right]\frac{\partial P_n^m(\theta_c)}{\partial\theta_c} \\ B_r &= \sum_{n=1}^{\infty}\left(\frac{r_0}{r}\right)^{n+2}(n+1)\sum_{m=0}^{n}\left[g_n^m\cos(m\phi) + h_n^m\sin(m\phi)\right]P_n^m(\theta_c) \end{aligned} \tag{25}$$

where $r_0 = 6371.2 \times 10^3$ km is the reference radius of the Earth, respectively.

The average current in the EDT is defined as

$$I_{ave} = \frac{1}{L}\int_0^L I(s)ds \tag{26}$$

The open-loop optimal control problem for EDT deorbit can be stated as finding a state-control pair $\{x(t), v(t)\}$ over each time interval $[t_i, t_{i+1}]$ to minimize a cost function of the negative work done by the electrodynamic force

$$J = \int_{t_i}^{t_{i+1}} \vec{F}_e \cdot \vec{v}\, dt \triangleq \int_{t_i}^{t_{i+1}} \Pi(x, v, t)dt \tag{27}$$

subject to the nonlinear state equations of libration motion

$$\begin{aligned} x_1' &= x_2 \\ x_2' &= 2\eta^3 e \sin v + 2(x_2 + \eta^2) x_4 \tan x_3 - 3\eta^3 \sin x_1 \cos x_1 \\ &\quad + [\sin i \tan x_3 (2 \sin u \cos x_1 - \cos u \sin x_1) - \cos i](\varsigma - \lambda) I_{ave} \frac{\mu_m}{\mu \tilde{m}} \eta^3 \end{aligned} \tag{28}$$

$$\begin{aligned} x_3' &= x_4 \\ x_4' &= -(x_2 + \eta^2)^2 \sin x_3 \cos x_3 - 3\eta^3 \cos^2 x_1 \sin x_3 \cos x_3 \\ &\quad - \sin i (2 \sin u \sin x_1 + \cos u \cos x_1)(\varsigma - \lambda) I_{ave} \frac{\mu_m}{\mu \tilde{m}} \eta^3 \end{aligned} \tag{29}$$

where $(x_1, x_2, x_3, x_4) = (\alpha, \alpha', \beta, , \beta')$, $\eta = 1 + e \cos v$, $\dot{v} = (\mu / p^3)^{0.5} (1 + e \cos v)^2$, $r = p(1 + e \cos v)^{-1}$, $\lambda = (m_1 + 0.5 m_t)/m_{EDT}$ is determined by the mass ratio between the end-bodies, and ς is determined by the distribution of current along the EDT, such that, $\varsigma = I_{ave}^{-1} L^{-2} \int_0^L sI(s)ds$. Accordingly, $\varsigma = 0.5$ is used for the assumption of a constant current in the EDT. The initial conditions $\boldsymbol{x}(t_i) = \boldsymbol{x}_{start}$ and the box constraint $|\alpha| \le \alpha_{\max}$, $|\beta| \le \beta_{\max}$, $I_{\min} \le I_{ave} \le I_{\max}$. The environmental perturbations are simplified by considering only the electrodynamic force with a simple non-tilted dipole model of geomagnetic field, such that,

$$\vec{\mathbf{B}} - \frac{\mu_m}{r^3} \cos u \sin i \, \vec{\boldsymbol{e}}_{ox} + \frac{\mu_m}{r^3} \cos i \, \vec{\boldsymbol{e}}_{oy} - \frac{2\mu_m}{r^3} \sin u \sin i \, \vec{\boldsymbol{e}}_{oz} \tag{30}$$

Accordingly, the electrodynamic force $\vec{\boldsymbol{F}}_e$ exerting on the EDT can be obtained as,

$$\vec{\mathbf{F}}_e = -\int_0^L \vec{\mathbf{B}} \times I \, \vec{\boldsymbol{l}} \, ds = -I_{ave} L \, \vec{\mathbf{B}} \times \vec{\boldsymbol{l}} \tag{31}$$

3.1.2. Results and discussion

The initial and boundary conditions of box constraints of the case are shown in **Tables 1** and **2**.

Parameters	Values
Mass of the main satellite	5 kg
Mass of subsatellite	1.75 kg
Mass of the tether	0.25 kg
Dimensions of main satellite	0.2 × 0.2 × 0.2 m
Dimensions of subsatellite	0.1 × 0.17 × 0.1 m
Tether length	500 m
Tether diameter	0.0005 m
Tether conductivity (aluminum)	$3.4014 \times 10^7 \ \Omega^{-1} \ \text{m}^{-1}$
Tether current lower/upper limits	0 ~ 0.8 A for the equatorial orbit
Orbital altitudes	700 ~ 800 km

Table 1. Parameters of an EDT system.

Parameters	Values
I_{max} (equatorial)	0.4 A
I_{max} (inclined)	$0.1\left[-\sin i \sin \overline{\beta}_B \sin\left(\Omega_G + \overline{\alpha}_B - \Omega\right) + \cos i \cos \overline{\beta}_B\right] \cos^{-1}\left(i - \overline{\beta}_B\right)$A
I_{min} (equatorial & inclined)	0 A
α_{max} (equatorial & inclined)	45 degrees
β_{max} (equatorial & inclined)	45 degrees

Table 2. Boundary Values of Box Constraints in the Open-Loop Trajectory Optimization.

Firstly, the validity of the proposed optimal control scheme in the equatorial orbit where the EDT system gets the highest efficiency is demonstrated. The solid line in **Figure 4** shows the time history of the EDT's average current control trajectory obtained from the open-loop optimal control problem. It is clearly shown that the average current in the open-loop case reaches the upper limit most of the time, which indicates the electrodynamic force being maximized for the fast deorbit. As expected, the current is not always at the upper limit in order to avoid the tumbling of the EDT system. This is evident that the timing of current reductions coincides with the peaks of pitch angles shown in **Figure 5**. The effectiveness of the proposed control scheme in terms of keeping libration stability is further demonstrated by the solid lines in **Figure 5**, where the trajectory of libration angles is no more than 45 degrees. It is also found that the amplitude of the pitch angle nearly reaches 45 degrees, the maximum allowed value, whereas the roll angle is very small in the whole deorbit process. As a comparison, tracking optimal control with the non-tilted dipole model and the IGRF 2000 model of the geomagnetic fields are conducted respectively.

The dashed lines in **Figures 3** and **4** show the tracking control simulations where all perturbations mentioned before are included with the non-tilted dipole geomagnetic field model. As

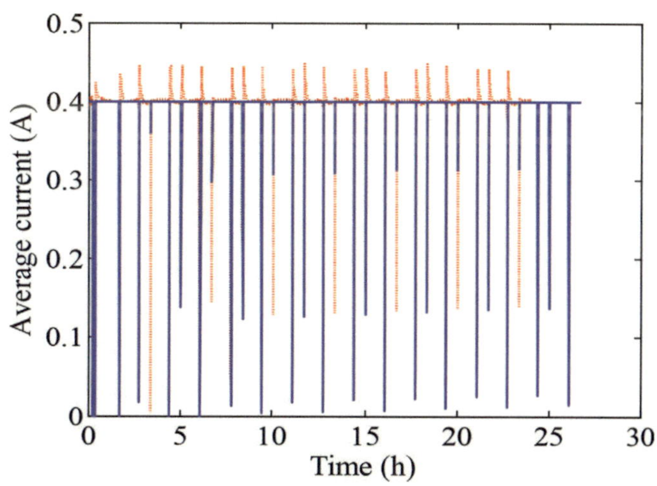

Figure 3. Time history of average current in the equatorial orbit (non-tilted dipole geomagnetic model). Solid line: Open-loop state trajectory. Dashed line: Close-loop tracking trajectory.

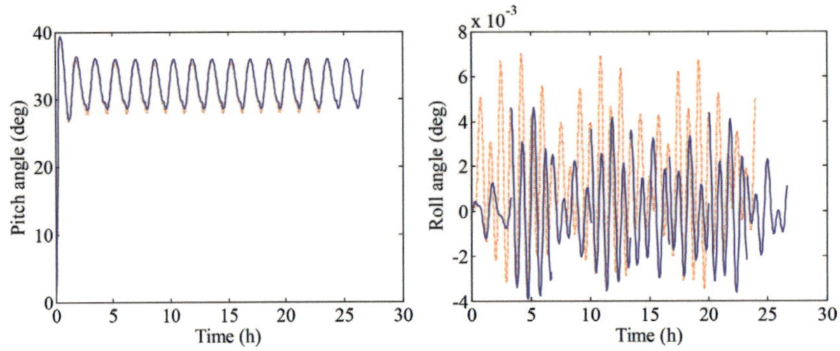

Figure 4. Time history of pitch and roll angles in the equatorial orbit (non-tilted dipole geomagnetic model). Solid line: Open-loop state trajectory. Dashed line: Close-loop tracking trajectory.

expected, the closed-loop tracking control works well in this case since the primary electrodynamic force perturbation is the same as the one used in the open-loop trajectory optimization. It is shown clearly in **Figure 4** that the pitch angle under the proposed closed-loop control tracks the open-loop optimal trajectory very closely with this simple environment model. **Figure 4** also shows the roll angle is almost zero even if it is not tracked. At the same time, **Figure 3** shows that the current control modification to the optimal current trajectory is relative small, i.e., 12% above the maximum current, for the same reason. Now the same cases are analyzed again using a more accurate geomagnetic field model – the IGRF 2000 model with up to 7th order terms (**Figures 5** and **6**). The solid line in **Figure 5** is the open-loop current control trajectory while the dashed line is the modified current control input obtained by the receding horizon control. Compared with **Figure 3**, it shows more current control modifications are

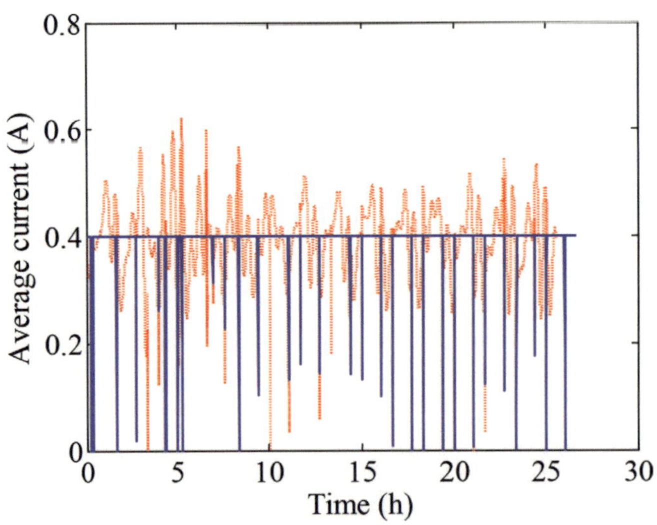

Figure 5. Time history of average current in the equatorial orbit (IGRF 2000 model). Solid line: Open-loop state trajectory. Dashed line: Close-loop tracking trajectory.

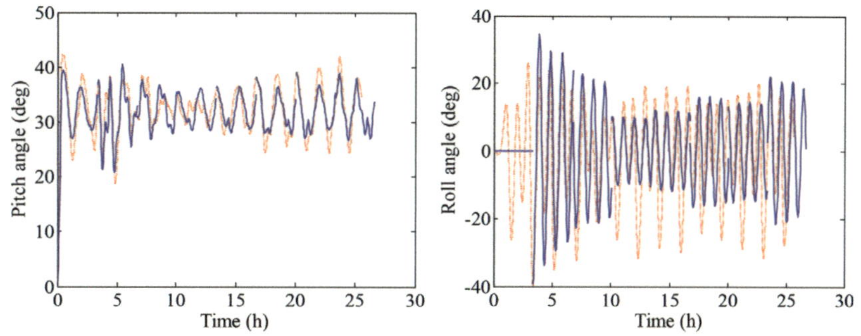

Figure 6. Time history of pitch and roll angles in the equatorial orbit (IGRF 2000 model). Solid line: Open-loop control trajectory. Dashed line: Close-loop tracking trajectory.

needed to track the open-loop control trajectory because of larger differences in dynamic models between the open-loop and closed-loop optimal controls, primarily due to the different geomagnetic field models. Because of the same reason, it is noticeable in **Figure 6** that the instantaneous states of the EDT system, controlled by the closed-loop optimal control law, are different from the open-loop reference state trajectory at the end of the interval. The instantaneous states are used for the next interval as initial conditions for the open-loop optimal control problem to derive the optimal control trajectory in that interval. This is reflected in **Figure 6** that the solid lines are discontinuous at the beginning of each interval. The dashed line in **Figure 7** shows that the pitch and roll angle under the closed-loop control has been controlled to the open-loop control trajectory, indicating the effectiveness of the proposed optimal control law. The roll angle is not controlled in this case as mentioned before. Compared **Figure 4** with **Figure 6**, it shows that the roll angle increases significantly since there is an out-of-plane component of the electrodynamic force resulting from the IGRF 2000 geomagnetic model. However, the amplitude of the roll angle is acceptable within the limits and will not lead to a tumbling of the EDT system.

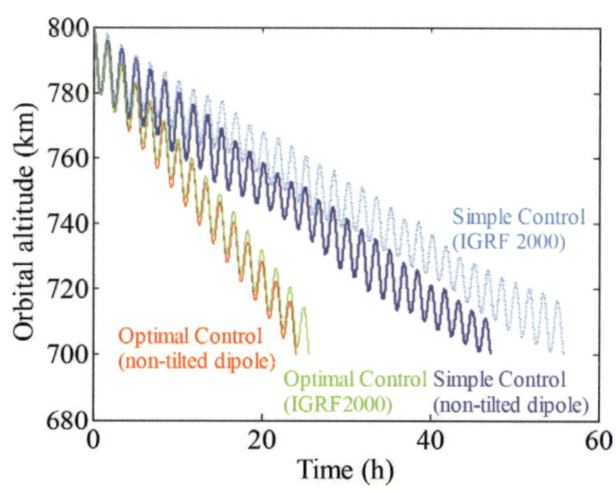

Figure 7. Comparison of EDT deorbit rates using different control laws and geomagnetic field models.

Finally, we make a comparison to show the performance of the proposed onboard parallel optimal control law from the aspect of deorbit rate. A simple current on–off control law from a previous work of Zhong and Zhu [21] is used here as baselines for the comparison of EDT deorbit efficiency. The current on–off control becomes active only if the libration angles exceed the maximum allowed values. Furthermore, it will turn on the current only in the condition that the electrodynamic force does negative work in both pitch and roll directions. In this paper, the maximum allowed amplitude for pitch and roll angles was set to 20° and the turned-on current was assumed to be 0.4 A, roughly the average value of the current control input into the closed-loop optimal control. Besides, a minimum interval of 10 minutes for the switching was imposed to avoid equipment failure that might happen due to the frequent switching. **Figure 8** shows the comparisons of the deorbit rates in different cases (the present optimal control and the current switching control with the non-tilted dipole or the IGRF 2000 model of geomagnetic field). It is shown that the EDT deorbit under the proposed optimal control scheme is faster than the current on–off control regardless which geomagnetic field model is used. The deorbit time of proposed optimal control based on the IGRF2000 model is about 25 hours, which equals approximately 15 orbits, whereas the deorbit time of simple current on–off control based on the same geomagnetic field model is about 55 hours, which equals approximately 33 orbits. The results also indicate that in the optimal control scheme, the effect is mostly shown in the current control input, instead of the deorbit rate, where **Figure 6** shows much more current control effort is required due to the different magnetic field models were used in the open-loop control trajectory optimization and the closed-loop optimal tracking control.

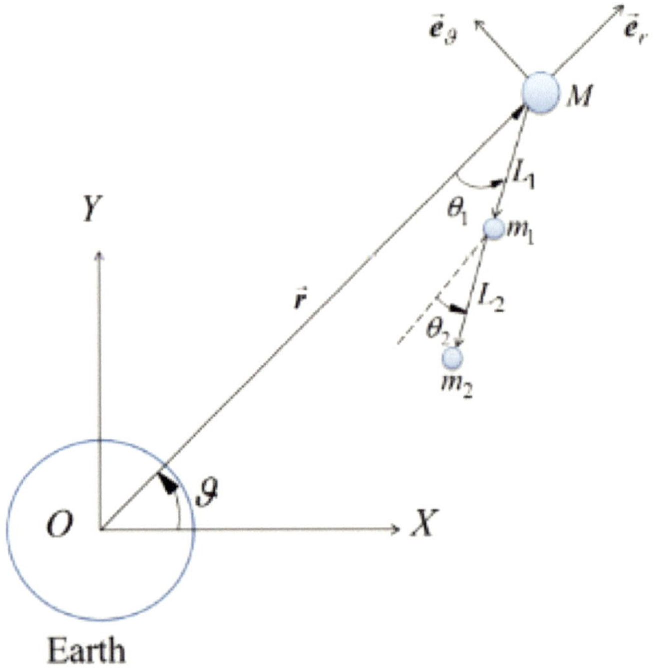

Figure 8. PSE system.

3.2. Parallel optimal algorithm in libration suppression of partial space elevator

For further test of the effect of the onboard parallel algorithm, the proposed control method is used to suppress the libration motions of the partial space elevator system. As studied in [24], this system is a non-equilibrium nonlinear dynamic system. It is difficult to suppress such a system in the mission period by using the common control design methods. In this case, we mainly concern obtaining the local time optimization.

3.2.1. *Problem formulation*

Consider an in-plane PSE system in a circular orbit is shown in **Figure 8**, where the main satellite, climber and the end body are connected by two inelastic tethers L_1 and L_2, respectively. The masses of the tethers are neglected. Assuming the system is subject to a central gravitational field and orbiting in the orbital plane. All other external perturbations are neglected. The main satellite, climber and end body are modeled as three point masses (M, m_1 and m_2) since the tether length is much greater than tethered bodies [5, 23]. Thus, the libration motions can be expressed in an Earth inertial coordinate system OXY with its origin at the centre of Earth. Denoting the position of the main satellite (M) by a vector $\mathbf{r}$ measuring from the centre of Earth. The climber m_1 is connected to the main satellite M by a tether 1 with the length of L_1 and a libration angle θ_1 measured from the vector $\mathbf{r}$. The distance between them is controlled by reeling in/out tether 1 at main satellite. The end body m_2 is connected to m_1 by a tether 2 with the length of L_2 and a libration angle θ_2 measured from the vector $\mathbf{r}$. The length of tether 2 L_2 is controlled by reeling in or out tether 2 at end body. The mass of the main satellite is assumed much greater than the masses of the climber and the end body. Therefore, the CM of the PSE system can be assumed residing in the main satellite that moves in a circular orbit. Based on the aforementioned assumptions, the dynamic equations can be written as.

$$\ddot{\theta}_1 = -\frac{3\omega^2 \sin 2\theta_1}{2} - \frac{2(\omega + \dot{\theta}_1)\dot{L}_1}{L_1} - \frac{\sin(\theta_1 - \theta_2)T_2}{L_1 m_1} \tag{32}$$

$$\ddot{\theta}_2 = \frac{-3\omega^2 \sin 2\theta_2}{2} - \frac{2(\omega + \dot{\theta}_2)(\dot{L}_c - \dot{L}_1)}{L_0 - L_1 + L_c} + \frac{\sin(\theta_1 - \theta_2)T_1}{(L_0 - L_1 + L_c)m_1} \tag{33}$$

$$\ddot{L}_1 = 3\omega^2 L_1 \cos^2\theta_1 + 2\omega L_1\dot{\theta}_1 + L_1\dot{\theta}_1^{\,2} - \frac{T_1}{m_1} + \frac{\cos(\theta_1 - \theta_2)T_2}{m_1} \tag{34}$$

$$\begin{aligned}\ddot{L}_c &= 3\omega^2(L_0 - L_1 + L_c)\cos^2\theta_2 + 3\omega^2 L_1 \cos^2\theta_1 + (2\omega + \dot{\theta}_2)(L_0 - L_1 + L_c)\dot{\theta}_2 \\ &\quad + (2\omega + \dot{\theta}_1)L_1\dot{\theta}_1 + \frac{[\cos(\theta_1 - \theta_2) - 1]T_1}{m_1} - \frac{[m_1 - m_2\cos(\theta_1 - \theta_2) + m_2]T_2}{m_1 m_2}\end{aligned} \tag{35}$$

where L_0 is the initial total length of two pieces of the tethers and L_c is the length increment relates to L_0.

The libration angles are required to be kept between the desired upper/lower bounds in the climber transfer process. The accessing process is divided into a series of intervals. In this case, we modified the aforementioned parallel optimal algorithm. The total transfer length $L_{10} - L_{1f}$

of the climber is discretized evenly. The optimal trajectory should be found to make the transfer time minimize in each equal tether transferring length. Then the cost function can be rewritten as

$$J_i = t_f \tag{36}$$

subject to the simplified dynamic equations

$$\ddot{\theta}_1 = -3\omega^2\theta_1 - \frac{2(\omega+\dot{\theta}_1)\dot{L}_1}{L_1} - \frac{(\theta_1-\theta_2)T_2}{L_1 m_1} \tag{37}$$

$$\ddot{\theta}_2 = -3\omega^2\theta_2 - \frac{2(\omega+\dot{\theta}_2)(\dot{L}_c-\dot{L}_1)}{L_0-L_1+L_c} + \frac{(\theta_1-\theta_2)T_1}{(L_0-L_1+L_c)m_1} \tag{38}$$

$$\ddot{L}_1 = 3\omega^2 L_1 + 2\omega L_1\dot{\theta}_1 + L_1\dot{\theta}_1^{\,2} + \frac{T_2-T_1}{m_1} \tag{39}$$

$$\ddot{L}_c = 3\omega^2(L_0+L_c) + (2\omega+\dot{\theta}_2)(L_0-L_1+L_c)\dot{\theta}_2 + (2\omega+\dot{\theta}_1)L_1\dot{\theta}_1 - \frac{T_2}{m_2} \tag{40}$$

where $\boldsymbol{u} = (T_1,\ T_2)$, $\boldsymbol{x} = \left(\theta_1, \dot{\theta}_1, \theta_2, \dot{\theta}_2, L_1, \dot{L}_1\right)$ and i denotes the interval number. In (26) the gravitational perturbations and the trigonometric functions are ignored, then they can be simplified following the assumptions: $\sin\theta_j \sim \theta_j,\ \ \cos\theta_j \sim 1\ \ (j = 1,\ 2)$. All the errors between simple model and the entire model are regarded as perturbations.

To ensure the availability and the suppression of the libration angles, following constrains are also required to be subjected $0 \le T_1 \le T_{1\max}, 0 \le T_2 \le T_{2\max}, |\theta_1| \le \theta_{1\max}, |\theta_2| \le \theta_{2\max}, |L_c| \le L_{cLimit}$ $\dot{L}_{1m} \le \dot{L}_1 \le \dot{L}_{1M}, \dot{L}_{cm} \le \dot{L}_c \le \dot{L}_{cM}$, where T_{1max} and T_{2max} are the upper bounds of the tension control inputs T_1 and T_2, respectively. θ_{1max}, θ_{2max} and L_{cLimit} are the magnitudes of libration angles θ_1, θ_2 and the maximum available length scale of L_c, respectively. $\dot{L}_{1m}$ and $\dot{L}_{1M}$ are the lower and upper bounds of climber's moving speed $\dot{L}_1$, respectively. $\dot{L}_{cm}$ and $\dot{L}_{cM}$ are the lower and upper bounds of end-bodies' moving speed $\dot{L}_c$, respectively. It should be noting that, to avoid the tether slacking, the control tensions are not allowed smaller than zero. Dividing the time interval $[t_i,\ t_{i+1}]$ evenly into n subintervals. The cost function minimization problem for each time interval can be solved by Hermite–Simpson method, due its simplicity and accuracy [17]. Then nonlinear programming problem is to search optimal values for the programming variables that minimize the cost function for each interval shown in (25). The closed-loop optimal tracking control method, is same as that in case 1. Direct transcription methods are routinely implemented with standard nonlinear programming (NLP) software. The sparse sequential quadratic programming software SNOPT is used via a MATLAB-executable file interface.

3.2.2. Results and discussion

The proposed control scheme is used to suppress the libration angles of the PSE system in the ascending process with following system parameters and initial conditions: r = 7100 *km*,

m_1 = 500 kg, m_2 = 1000 kg, $\theta_1(0)$ = $\theta_2(0)$ = 0, L_0 = 20 km, $L_1(0)$ = 19,500 m, L_c (0) = 0, $\dot{\theta}_1(0) = \dot{\theta}_2(0) = 0$, $\dot{L}_1(0) = -20$ m/s, and $\dot{L}_c(0) = 0$ for the ascending process. The whole transfer trajectory is divided into 50 intervals. The climber's ascending speed along tether 1 is allowed to be controlled to help suppress the libration angles and keep the states of the system in an acceptable area. The constrains are set as $T_{1\max} = T_{2\max} = 200N$, $\theta_{1\max} = \theta_{2\max} = 0.3$ rad, $\dot{L}_{1m} = -15$ m/s, $\dot{L}_{1M} = -25$ m/s, $\dot{L}_{cm} = -10$ m/s and $\dot{L}_{cM} = 10$ m/s.

The simulation results of this case are shown in **Figures 9–11**. The climber's open-loop libration angle approaches its upper bound at 850 s. After 850 s θ_1 is kept at 0.3 rad by the end of the ascending period, see the dashed line in **Figure 10**. Using the closed-loop control, the tracking trajectory of θ_1 matches the open-loop trajectory very well overall, see solid line in **Figure 9**. A short gap appears between 875 s – 880 s, this is caused by the errors of the model and computation. **Figure 9** also shows the changes of the trajectories of θ_2.The trajectory of θ_2 obtained by closed-loop control tracks the open-loop trajectory well and reaches 0.1 rad by the end of the ascending period. The closed-loop trajectories of L_1 and L_c are shown in **Figure 10**. They are the reflections of the control inputs. Both L_1 and L_c show smooth fluctuations between 40s and 350 s. **Figure 10** shows the time history of trajectories of $\dot{L}_1$ and $\dot{L}_c$,

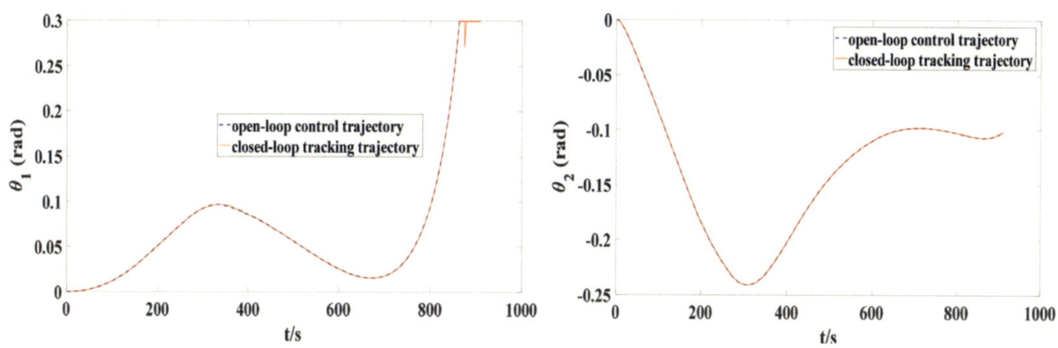

Figure 9. Libration angle of θ_1 and θ_2 with variable climber speed.

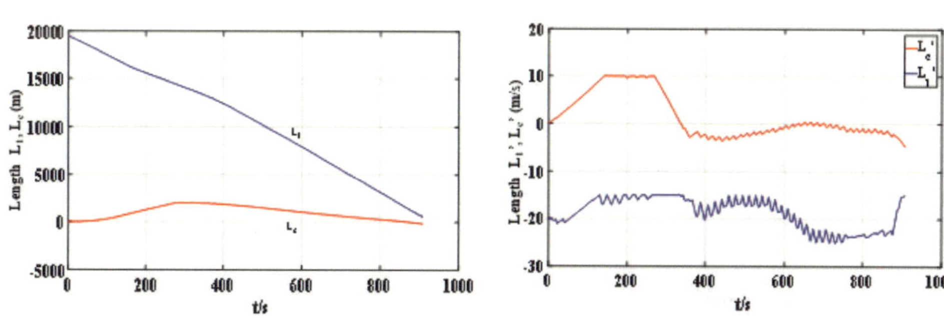

Figure 10. Length and its changing ratio of L_1, L_c with variable climber speed.

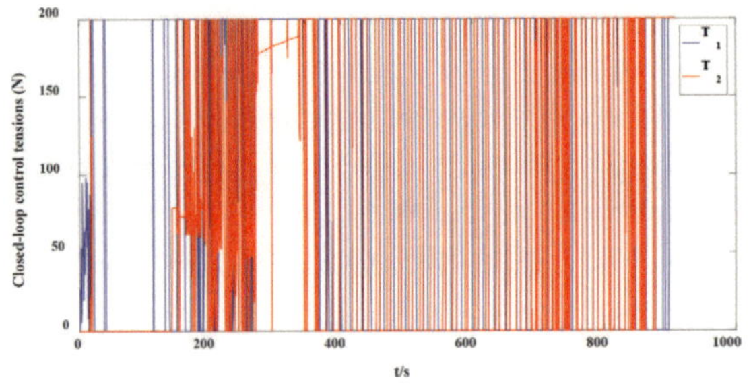

Figure 11. Control inputs for the closed-loop control with changing climber speed.

respectively. In the first 140 s the trajectory of $\dot{L}_c$ increases continuously until reaches its upper bound. Then, it keeps at 10 *m/s* by 270 s with some slight fluctuations. From 270 s to 355 s, it reduces continuously to −3 m/s. After that, $\dot{L}_c$ fluctuates around −3 m/s by the end of the transfer period. As a reflection of the control input, $\dot{L}_1$ also shows fluctuation during the transfer phase with obvious small-scale fluctuations appear in the period of 120 s – 260 s and 360 s – 750 s. This impacts during the whole transfer period, the changeable speed of the climber has the ability to help the suppression of the libration angles and states trajectory tracking. This time history of the control inputs is shown in **Figure 11** with frequent changes between its lower bound and upper bound.

4. Conclusions

This chapter investigated a piecewise parallel onboard optimal control algorithm to solve the optimal control issues in complex nonlinear dynamic systems in aerospace engineering. To test the validity of the proposed two-phase optimal control scheme, the long-term tether libration stability and fast nano-satellite deorbit under complex environmental perturbations and the libration suppression for PSE system are considered. For EDT system, instead of optimizing the control of fast and stable nano-satellite deorbit over the complete process, the current approach divides the deorbit process into a set of intervals. For the PSE system, each time interval is set depends on the minimize transfer time for equal transfer length interval. Within each interval, the predicting phase simplifies significantly the optimal control problem. The dynamic equations of libration motion are further simplified to reduce computational loads using the simple dynamic models. The trajectory of the stable libration states and current control input is then optimized for the fast deorbit within the interval based on the simplified dynamic equations. The tracking optimizes the trajectory tracking control using the finite receding horizon control theory within the time interval corresponding to the open-loop control state trajectory with the same interval number. By applying the close-loop control modification, the system motions are integrated without any simplification of the dynamics or environmental perturbations and the instantaneous states of the orbital and libration motions. The *i*-th time interval's closed-loop tracking is processed in tracking phase while the

(i + 1)-th time interval's optimal state trajectory is predicted in the predicting phase. This prediction is based on the error between the real state and the predicting state in the (i-1)-th time interval. By repeating the process, the optimal control problem can be achieved in a piecewise way with dramatically reduced computation effort. Compared with the current on–off control where the stable libration motion is the only control target, numerical results show that the proposed optimal control scheme works well in keeping the libration angles within an allowed limit.

Acknowledgements

This work is funded by the Discovery Grant of the Natural Sciences and Engineering Research Council of Canada, the National Natural Science Foundation of China, Grand No. 11472213 and the Chinese Scholarship Council Scholarship No. 201606290135.

A. Appendix

The detailed expressions for the matrixes $\boldsymbol{C}$ and $\boldsymbol{M}$ are shown as followed,

$$\boldsymbol{C} = [\boldsymbol{C}_1 \;\; \boldsymbol{C}_2 \;\; \boldsymbol{C}_3], \qquad \boldsymbol{M} = \begin{bmatrix} \boldsymbol{M}_{11} & \boldsymbol{M}_{12} & 0 \\ \boldsymbol{M}_{12}^T & \boldsymbol{M}_{22} & 0 \\ 0 & 0 & \boldsymbol{M}_{33} \end{bmatrix}$$

$$\boldsymbol{C}_1 = \begin{bmatrix} \boldsymbol{E} & & & & 0 \\ \boldsymbol{x}_0^1 & \boldsymbol{x}_0^2 & & & \\ & \boldsymbol{x}_1^1 & \boldsymbol{x}_1^2 & & \\ & & \ddots & \ddots & \\ & & & \boldsymbol{x}_{n-1}^1 & \boldsymbol{x}_{n-1}^2 \\ 0 & & & & \boldsymbol{E} \end{bmatrix}, \quad \boldsymbol{C}_2 = \begin{bmatrix} 0 & & & & 0 \\ \boldsymbol{x}_0^3 & \boldsymbol{x}_0^4 & & & \\ & \boldsymbol{x}_1^3 & \boldsymbol{x}_1^4 & & \\ & & \ddots & \ddots & \\ & & & \boldsymbol{x}_{n-1}^3 & \boldsymbol{x}_{n-1}^4 \\ 0 & & & & 0 \end{bmatrix}, \boldsymbol{C}_3 = \frac{2}{3}\overline{\gamma} \begin{bmatrix} 0 & & 0 \\ \boldsymbol{B}_{0.5} & & \\ & \ddots & \\ & & \boldsymbol{B}_{n-0.5} \\ 0 & & 0 \end{bmatrix}$$

$$\boldsymbol{x}_j^1 = \boldsymbol{E} + \frac{\overline{\gamma}}{6}\boldsymbol{A}_j + \frac{\overline{\gamma}}{3}\boldsymbol{A}_{j+0.5} + \frac{\overline{\gamma}^2}{12}\boldsymbol{A}_{j+0.5}\boldsymbol{A}_j, \qquad \boldsymbol{x}_j^2 = -\boldsymbol{E} + \frac{\overline{\gamma}}{6}\boldsymbol{A}_{j+1} + \frac{\overline{\gamma}}{3}\boldsymbol{A}_{j+0.5} - \frac{\overline{\gamma}^2}{12}\boldsymbol{A}_{j+0.5}\boldsymbol{A}_{j+1}$$

$$\boldsymbol{x}_j^3 = \frac{\overline{\gamma}^2}{12}\boldsymbol{A}_{j+0.5}\boldsymbol{B}_j + \frac{\overline{\gamma}}{6}\boldsymbol{B}_j, \qquad \boldsymbol{x}_j^4 = -\frac{\overline{\gamma}^2}{12}\boldsymbol{A}_{j+0.5}\boldsymbol{B}_{j+1} + \frac{\overline{\gamma}}{6}\boldsymbol{B}_{j+1}$$

$$
\boldsymbol{M}_{11} = \begin{bmatrix} \varpi_{00}^{11} & \varpi_{01}^{11} & & & 0 \\ \left(\varpi_{01}^{11}\right)^T & \varpi_{11}^{11} & & & \\ & & \ddots & & \\ & & & \varpi_{n-1n-1}^{11} & \varpi_{n-1n}^{11} \\ 0 & & & \left(\varpi_{n-1n}^{11}\right)^T & \varpi_{nn}^{11} \end{bmatrix} \quad \boldsymbol{M}_{22} = \begin{bmatrix} \varpi_{00}^{22} & \varpi_{01}^{22} & & & 0 \\ \left(\varpi_{01}^{22}\right)^T & \varpi_{11}^{22} & & & \\ & & \ddots & & \\ & & & \varpi_{n-1n-1}^{22} & \varpi_{n-1n}^{22} \\ 0 & & & \left(\varpi_{n-1n}^{22}\right)^T & \varpi_{nn}^{22} \end{bmatrix}
$$

$$
\boldsymbol{M}_{12} = \begin{bmatrix} \varpi_{00}^{12} & \varpi_{01}^{12} & & & 0 \\ \varpi_{10}^{12} & \varpi_{11}^{12} & & & \\ & & \ddots & & \\ & & & \varpi_{n-1n-1}^{12} & \varpi_{n-1n}^{12} \\ 0 & & & \varpi_{nn-1}^{12} & \varpi_{nn}^{12} \end{bmatrix} \qquad \boldsymbol{M}_{33} = \frac{2}{3}\overline{\gamma} \begin{bmatrix} R & & 0 \\ & \ddots & \\ 0 & & R \end{bmatrix}
$$

$$
\varpi_{jj}^{11} = \frac{\overline{\gamma}}{6}\left(4\boldsymbol{Q} + \frac{\overline{\gamma}^2}{8}\boldsymbol{A}_j^T\boldsymbol{Q}\boldsymbol{A}_j\right), \qquad \varpi_{jj+1}^{11} = \frac{\overline{\gamma}}{6}\left(\boldsymbol{Q} - \frac{\overline{\gamma}^2}{16}\boldsymbol{A}_j^T\boldsymbol{Q}\boldsymbol{A}_{j+1} + \frac{\overline{\gamma}}{4}\boldsymbol{A}_j^T\boldsymbol{Q} - \frac{\overline{\gamma}}{4}\boldsymbol{Q}\boldsymbol{A}_{j+1}\right)
$$

$$
\varpi_{00}^{11} = \frac{\overline{\gamma}}{6}\left(2\boldsymbol{Q} + \frac{\overline{\gamma}^2}{16}\boldsymbol{A}_0^T\boldsymbol{Q}\boldsymbol{A}_0 + \frac{\overline{\gamma}}{4}\boldsymbol{A}_0^T\boldsymbol{Q} + \frac{\overline{\gamma}}{4}\boldsymbol{Q}\boldsymbol{A}_0\right), \quad \varpi_{nn}^{11} = \frac{\overline{\gamma}}{6}\left(2\boldsymbol{Q} + \frac{\overline{\gamma}^2}{16}\boldsymbol{A}_n^T\boldsymbol{Q}\boldsymbol{A}_n - \frac{\overline{\gamma}}{4}\boldsymbol{A}_n^T\boldsymbol{Q} - \frac{\overline{\gamma}}{4}\boldsymbol{Q}\boldsymbol{A}_n\right) + S_f
$$

$$
\varpi_{jj}^{22} = \frac{\overline{\gamma}}{6}\left(2R + \frac{\overline{\gamma}^2}{8}\boldsymbol{B}_j^T\boldsymbol{Q}\boldsymbol{B}_j\right), \qquad \varpi_{jj+1}^{22} = -\overline{\gamma}\boldsymbol{B}_j^T\boldsymbol{Q}\boldsymbol{B}_{j+1}
$$

$$
\varpi_{00}^{22} = \frac{\overline{\gamma}}{6}\left(R + \frac{\overline{\gamma}^2}{16}\boldsymbol{B}_0^T\boldsymbol{Q}\boldsymbol{B}_0\right), \qquad \varpi_{nn}^{22} = \frac{\overline{\gamma}}{6}\left(R + \frac{\overline{\gamma}^2}{16}\boldsymbol{B}_n^T\boldsymbol{Q}\boldsymbol{B}_n\right)
$$

$$
\varpi_{jj}^{12} = \frac{\overline{\gamma}^3}{48}\boldsymbol{A}_j^T\boldsymbol{Q}\boldsymbol{B}_j, \qquad \varpi_{jj+1}^{12} = -\frac{\overline{\gamma}^2}{24}\left(\boldsymbol{Q}\boldsymbol{B}_{j+1} + \frac{\overline{\gamma}}{4}\boldsymbol{A}_j^T\boldsymbol{Q}\boldsymbol{B}_{j+1}\right)
$$

$$
\varpi_{j+1j}^{12} = \frac{\overline{\gamma}^2}{24}\left(\boldsymbol{Q}\boldsymbol{B}_j - \frac{\overline{\gamma}}{4}\boldsymbol{A}_{j+1}^T\boldsymbol{Q}\boldsymbol{B}_j\right), \quad \varpi_{00}^{12} = \frac{\overline{\gamma}^2}{24}\left(\boldsymbol{Q}\boldsymbol{B}_0 + \frac{\overline{\gamma}}{4}\boldsymbol{A}_0^T\boldsymbol{Q}\boldsymbol{B}_0\right), \quad \varpi_{nn}^{12} = -\frac{\overline{\gamma}^2}{24}\left(\boldsymbol{Q}\boldsymbol{B}_{\overline{N}} - \frac{\overline{\gamma}}{4}\boldsymbol{A}_n^T\boldsymbol{Q}\boldsymbol{B}_n\right)
$$

where E is the unit matrix which has the same dimension as A_j, and $j = 0,1,2,\ldots,n-1$.

Author details

Zheng Hong Zhu[1*] and Gefei Shi[1,2,3]

*Address all correspondence to: gzhu@yorku.ca

1 Department of Mechanical Engineering, York University, Toronto, Ontario, Canada

2 National Key Laboratory of Aerospace Flight Dynamics, Northwestern Polytechnical University, Xi'an, PR China

3 School of Astronautics, Northwestern Polytechnical University, Xi'an, PR China

References

[1] Zhong R, Zhu ZH. Optimal control of Nanosatellite fast Deorbit using Electrodynamic tether. Journal of Guidance Control and Dynamics. 2014;**37**:1182-1194. DOI: 10.2514/1.62154

[2] Jablonski AM, Scott R. Deorbiting of microsatellites in low earth orbit (LEO) – An introduction. Canadian Aeronautics and Space Journal. 2009;**55**:55-67

[3] Stevens RE, Baker WP. Optimal control of a librating Electrodynamic tether performing a multirevolution orbit change. Journal of Guidance Control and Dynamics. 2009;**32**: 1497-1507. DOI: 10.2514/1.42679

[4] Williams P. Optimal control of Electrodynamic tether orbit transfers using timescale separation. Journal of Guidance Control and Dynamics. 2010;**33**:88-98. DOI: 10.2514/1.45250

[5] Aslanov V, Ledkov A. Dynamics of Tethered Satellite Systems. Cambridge, UK: Elsevier; 2012

[6] Wen H, Zhu ZH, Jin DP, Hu HY. Space tether deployment control with explicit tension constraint and saturation function. Journal of Guidance, Control, and Dynamics. 2016;**39**: 916-921. DOI: 10.2514/1.G001356

[7] Ma ZQ, Sun GH, Li ZK. Dynamic adaptive saturated sliding mode control for deployment of tethered satellite system. Aerospace Science and Technology. 2017;**66**:355-365. DOI: 10.1016/j.ast.2017.01.030

[8] Jin DP, Hu HY. Optimal control of a tethered subsatellite of three degrees of freedom. Nonlinear Dynamics. 2006;**46**:161-178. DOI: 10.1007/s11071-006-9021-4

[9] Williams P. Deployment/retrieval optimization for flexible tethered satellite systems. Nonlinear Dynamics. 2008;**52**:159-179. DOI: 10.1007%2Fs11071-007-9269-3

[10] Williams P, Ockels W. Climber motion optimization for the tethered space elevator. Acta Astronautica. 2010;**66**:1458-1467. DOI: 10.1016/j.actaastro.2009.11.003

[11] Kojima H, Fukatsu K, Trivailo PM. Mission-function control of tethered satellite/climber system. Acta Astronautica. 2015;**106**:24-32. DOI: 10.1016/j.actaastro.2014.10.024

[12] Williams P. Optimal control of Electrodynamic tether orbit transfers using timescale separation. Journal of Guidance Control and Dynamics. 2010;**33**:88-98. DOI: 10.2514/1.45250

[13] Betts JT. Practical methods for optimal control using nonlinear programming. Society for Industrial and Applied Mathematics: Advances in Control and Design Series, Philadelphia, PA. 2001:65-72

[14] Zhong R, Xu S. Orbit-transfer control for TSS using direct collocation method. Acta Aeronautica et Astronautica Sinica. 2010;**31**:572-578

[15] Bryson AE, Ho YC. Chapter 2. In: Applied Optimal Control. New York: Hemisphere; 1975

[16] Herman AL, Conway BA. Direct optimization using collocation based on high-order gauss-Lobatto Quadrature rules. Journal of Guidance Control and Dynamics. 1996;**19**: 592-599. DOI: 10.2514/3.21662

[17] Hargraves CR, Paris SW. Direct trajectory optimization using nonlinear programming and collocation. Journal of Guidance, Control, and Dynamics. 1987;**10**:338-342. DOI: 10.2514/3.20223

[18] Vallado DA. Fundamentals of Astrodynamics and Applications. New York, Springer: Microcosm Press; 2007

[19] NASA, Earth Fact Sheet, http://nssdc.gsfc.nasa.gov/planetary/factsheet/earthfact.html, [retrieved March 16, 2012]

[20] Davis J. Technical note. In: Mathematical Modeling of Earth's Magnetic Field. Blacksburg: Virginia Tech; 2004

[21] Zhong R, Zhu ZH. Long-term libration dynamics and stability analysis of Electrodynamic tethers in spacecraft Deorbit. Journal of Aerospace Engineering. 2012. DOI: 10.1061/(ASCE)AS.1943-5525.0000310

[22] Shi G, Zhu Z, Zhu ZH. Libration suppression of tethered space system with a moving climber in circular orbit. Nonlinear Dynamics. 2017:1-15. DOI: 10.1007/s11071-017-3919-x

[23] Wen H, Zhu ZH, Jin DP, Hu H. Constrained tension control of a tethered space-tug system with only length measurement. Acta Astronautica. 2016;**119**:110-117. DOI: doi.org/10.1016/j.actaastro.2015.11.011

[24] Fletcher R. Chapter 10. In: Practical Methods of Optimization. 2nd ed. New York: Wiley; 1989

Chapter 7

Bilevel Disjunctive Optimization on Affine Manifolds

Constantin Udriste, Henri Bonnel, Ionel Tevy and
Ali Sapeeh Rasheed

Additional information is available at the end of the chapter

http://dx.doi.org/10.5772/intechopen.75643

Abstract

Bilevel optimization is a special kind of optimization where one problem is embedded within another. The outer optimization task is commonly referred to as the upper-level optimization task, and the inner optimization task is commonly referred to as the lower-level optimization task. These problems involve two kinds of variables: upper-level variables and lower-level variables. Bilevel optimization was first realized in the field of game theory by a German economist von Stackelberg who published a book (1934) that described this hierarchical problem. Now the bilevel optimization problems are commonly found in a number of real-world problems: transportation, economics, decision science, business, engineering, and so on. In this chapter, we provide a general formulation for bilevel disjunctive optimization problem on affine manifolds. These problems contain two levels of optimization tasks where one optimization task is nested within the other. The outer optimization problem is commonly referred to as the leaders (upper level) optimization problem and the inner optimization problem is known as the followers (or lower level) optimization problem. The two levels have their own objectives and constraints. Topics affine convex functions, optimizations with auto-parallel restrictions, affine convexity of posynomial functions, bilevel disjunctive problem and algorithm, models of bilevel disjunctive programming problems, and properties of minimum functions.

Keywords: convex programming, affine manifolds, optimization along curves, bilevel disjunctive optimization, minimum functions

Mathematics Subject Classification 2010: 90C25 90C29, 90C30

1. Affine convex functions

In optimization problems [16, 17, 19, 23–27], one can use an *affine manifold* as a pair (M, Γ), where M is a smooth real n-dimensional manifold, and Γ is an affine symmetric connection on M. The connection Γ produces auto-parallel curves $x(t)$ via ODE system

$$\ddot{x}^h(t) + \Gamma^h_{ij}(x(t))\dot{x}^i(t)\dot{x}^j(t) = 0.$$

They are used for defining the convexity of subsets in M and convexity of functions $f : D \subset M \to \mathbb{R}$ (see also [3, 6]).

Definition 1.1 *An affine manifold (M, Γ) is called autoparallely complete if any auto-parallel $x(t)$ starting at $p \in M$ is defined for all values of the parameter $t \in \mathbb{R}$.*

Theorem 1.1 [1] *Let M be a (Hausdorff, connected, smooth) compact n-manifold endowed with an affine connection Γ and let $p \in M$. If the holonomy group $Hol_p(\Gamma)$ (regarded as a subgroup of the group $Gl(T_pM)$ of all the linear automorphisms of the tangent space T_pM) has compact closure, then (M, Γ) is autoparallely complete.*

Let (M, Γ) be an auto-parallely complete affine manifold. For a C^2 function $f : M \to \mathbb{R}$, we define the tensor $Hess_\Gamma f$ of components

$$(Hess_\Gamma f)_{ij} = \frac{\partial^2 f}{\partial x^i \partial x^j} - \Gamma^h_{ij}\frac{\partial f}{\partial x^h}.$$

Definition 1.2 *A C^2 function $f : M \to \mathbb{R}$ is called:*

(1) linear affine with respect to Γ if $Hess_\Gamma f = 0$, throughout;

(2) affine convex (convex with respect to Γ) if $Hess_\Gamma f \succcurlyeq 0$ (positive semidefinite), throughout.

The function f is: (1) linear affine if its restriction $f(x(t))$ on each autoparallel $x(t)$ satisfies $f(x(t)) = at + b$, for some numbers a, b that may depend on $x(t)$; (2) affine convex if its restriction $f(x(t))$ is convex on each auto-parallel $x(t)$.

Theorem 1.2 *If there exists a linear affine nonconstant function f on (M, Γ), then the curvature tensor field $R^h{}_{ikj}$ is in* $\operatorname{Ker} df$.

Proof. For given Γ, if we consider

$$\frac{\partial^2 f}{\partial x^i \partial x^j} = \Gamma^h_{ij}\frac{\partial f}{\partial x^h}$$

as a PDEs system (a particular case of a Frobenius-Mayer system of PDEs) with $\frac{1}{2}n(n+1)$ equations and the unknown function f, then we need the complete integrability conditions

$$\frac{\partial^3 f}{\partial x^i \partial x^j \partial x^k} = \frac{\partial^3 f}{\partial x^k \partial x^i \partial x^j}.$$

Since,

$$\frac{\partial^3 f}{\partial x^i \partial x^j \partial x^k} = \left(\frac{\partial \Gamma^h_{ij}}{\partial x^k} + \Gamma^l_{ij} \Gamma^h_{kl} \right) \frac{\partial f}{\partial x^h},$$

it follows

$$\frac{\partial f}{\partial x^h} R^h{}_{ikj} = 0, R^h{}_{ikj} = \frac{\partial \Gamma^h_{ij}}{\partial x^k} - \frac{\partial \Gamma^h_{ki}}{\partial x^j} + \Gamma^l_{ij} \Gamma^h_{kl} - \Gamma^l_{ki} \Gamma^h_{jl}.$$

Corollary 1.1 *If there exists n linear affine functions f_l, $l = 1, \ldots, n$ on (M, Γ), whose df_l are linearly independent, then Γ is flat, that is, $R^h{}_{ikj} = 0$.*

Of course this only means the curvature tensor is zero on the topologically trivial region we used to set up our co-vector fields $df_l(x)$. But we can always cover any manifold by an atlas of topologically trivial regions, so this allows us to deduce that the curvature tensor vanishes throughout the manifold.

Remark 1.1 *There is actually no need to extend $df_l(x)$ to the entire manifold. If this could be done, then $df_l(x)$ would now be everywhere nonzero co-vector fields; but there are topologies, for example, S^2, for which we know such things do not exist. Therefore, there are topological manifolds for which we are forced to work on topologically trivial regions.*

The following theorem is well-known [16, 17, 19, 23]. Due to its importance, now we offer new proofs (based on catastrophe theory, decomposing a tensor into a specific product, and using slackness variables).

Theorem 1.3 *Let $f : M \to \mathbb{R}$ be a C^2 function.*

(1) If f is regular or has only one minimum point, then there exists a connection Γ such that f is affine convex.

(2) If f has a maximum point x_0, then there is no connection Γ making f affine convex throughout.

Proof. For the Hessian $(Hess_\Gamma f)_{ij}$ be positive semidefinite, we need n conditions like inequalities and equalities. The number of unknowns Γ^h_{ij} is $\frac{1}{2}n^2(n+1)$. The inequalities can be replaced by equalities using slackness variables.

The first central idea for the proof is to use the catastrophe theory, since almost all families $f(x,c)$, $x=(x^1,\ldots,x^n)\in\mathbb{R}^n$, $c=(c_1,\ldots,c_m)\in\mathbb{R}^m$, of real differentiable functions, with $m\leq 4$ parameters, are structurally stable and are equivalent, in the vicinity of any point, with one of the following forms [15]:

We eliminate the case with maximum point, that is., Morse 0-saddle and the saddle point. Around each critical point (in a chart), the canonical form $f(x,c)$ is affine convex, with respect to appropriate locally defined linear connections that can be found easily. Using change of coordinates and the partition of unity, we glue all these connections to a global one, making $f(x,c)$ affine convex on M.

At any critical point x_0, the affine Hessian $Hess_\Gamma f$ is reduced to Euclidean Hessian, $\frac{\partial^2 f}{\partial x^i \partial x^j}(x_0)$. Then the maximum point condition or the saddle condition is contradictory to affine convexity condition.

A direct proof based on decomposition of a tensor: Let (M,Γ) be an affine manifold and $f: M\to\mathbb{R}$ be a C^2 function.

Suppose f has no critical points (is regular). If the function f is not convex with respect to Γ, we look to find a new connection $\overline{\Gamma}^h_{ij}=\Gamma^h_{ij}+T^h_{ij}$, with the unknown a tensor field T^h_{ij}, such that

$$\frac{\partial^2 f}{\partial x^i \partial x^j}(x)-\overline{\Gamma}^h_{ij}(x)\frac{\partial f}{\partial x^h}(x)=\sigma_{ij}(x), x\in M,$$

where $\sigma_{ij}(x)$ is a positive semi-definite tensor. A very particular solution is the decomposition $T^h_{ij}(x)=a^h(x)b_{ij}(x)$, where the vector field a has the property

$$D_a f=a^h(x)\frac{\partial f}{\partial x^h}(x)\neq 0, x\in M$$

and the tensor b_{ij} is

$$b_{ij}(x)=\frac{1}{D_a f}\left(\frac{\partial^2 f}{\partial x^i \partial x^j}(x)-\Gamma^h_{ij}(x)\frac{\partial f}{\partial x^h}(x)-\sigma_{ij}(x)\right), x\in M.$$

Remark 1.2 *The connection $\overline{\Gamma}^h_{ij}$ is strongly dependent on both the function f and the tensor field σ_{ij}.*

Suppose f has a minimum point x_0. In this case, observe that we must have the condition $\sigma_{ij}(x_0)=\frac{\partial^2 f}{\partial x^i \partial x^j}(x_0)$. Can we make the previous reason for $x\neq x_0$ and then extend the obtained connection by continuity? The answer is generally negative. Indeed, let us compute

$$b_{ij}(x_0)=\lim_{x\to x_0}\frac{1}{D_a f}\left(\frac{\partial^2 f}{\partial x^i \partial x^j}(x)-\Gamma^h_{ij}(x)\frac{\partial f}{\partial x^h}(x)-\sigma_{ij}(x)\right).$$

Here we cannot plug in the point x_0 because we get $\frac{0}{0}$, an indeterminate form.

To contradict, we fix an auto-parallel $\gamma(t), t \in [0, \epsilon)$, starting from minimum point $x_0 = \gamma(0)$, tangent to $\dot{\gamma}(0) = v$ and we compute (via l'Hôpital rule)

$$b_{ij}(x_0; v) = \lim_{t \to 0} b_{ij}(\gamma(t)) = \frac{\left(\frac{\partial^3 f}{\partial x^i \partial x^j \partial x^k}(x_0) - \Gamma^h_{ij}(x_0)\frac{\partial^2 f}{\partial x^h \partial x^k}(x_0) - \frac{\partial \sigma_{ij}}{\partial x^k}(x_0)\right) v^k}{a^h(x_0)\frac{\partial^2 f}{\partial x^h \partial x^k}(x_0) v^k}.$$

But this result depends on the direction v (different values along two different auto-parallels).

In some particular cases, we can eliminate the dependence on the vector v. For example, the conditions

$$\frac{\partial^3 f}{\partial x^i \partial x^j \partial x^l}(x_0) - \Gamma^h_{ij}(x_0)\frac{\partial^2 f}{\partial x^h \partial x^l}(x_0) - \frac{\partial \sigma_{ij}}{\partial x^l}(x_0)$$

$$= \rho\left(\frac{\partial^3 f}{\partial x^i \partial x^j \partial x^k}(x_0) - \Gamma^h_{ij}(x_0)\frac{\partial^2 f}{\partial x^h \partial x^k}(x_0) - \frac{\partial \sigma_{ij}}{\partial x^k}(x_0)\right),$$

$$a^h(x_0)\frac{\partial^2 f}{\partial x^h \partial x^l}(x_0) = \rho\left(a^h(x_0)\frac{\partial^2 f}{\partial x^h \partial x^k}(x_0)\right)$$

are sufficient to do this.

A particular condition for independence on v is

$$\frac{\partial^3 f}{\partial x^i \partial x^j \partial x^k}(x_0) - \Gamma^h_{ij}(x_0)\frac{\partial^2 f}{\partial x^h \partial x^k}(x_0) - \frac{\partial \sigma_{ij}}{\partial x^k}(x_0) = 0.$$

In this particular condition, we can show that we can build connections of previous type good everywhere.

1.1. Lightning through examples

Let us lightning our previous statements by the following examples.

Example 1.1 *(for the first part of the theorem) Let us consider the function* $f : \mathbf{R}^2 \to \mathbf{R}, f(x, y) = x^3 + y^3 + 3x + 3y$ *and* $\Gamma^h_{ij} = 0, i, j, h = 1, 2$. *Then* $\frac{\partial f}{\partial x} = 3x^2 + 3$, $\frac{\partial f}{\partial y} = 3y^2 + 3$ *and* f *has no critical point. Moreover, the Euclidean Hessian of* f *is not positive semi-definite overall. Let us make the above construction for* $\sigma_{ij}(x, y) = \delta_{ij}$. *Taking* $a^1 = a^2 = 1$, *we obtain the connection*

$$\overline{\Gamma}^h_{11} = \frac{6x - 1}{3x^2 + 3y^2 + 2}, \overline{\Gamma}^h_{22} = \frac{6y - 1}{3x^2 + 3y^2 + 2}, \overline{\Gamma}^h_{12} = \overline{\Gamma}^h_{21} = 0, \ h = 1, 2,$$

that is not unique.

Example 1.2 *(for one minimum point) Let us consider the function* $f : \mathbf{R}^2 \to \mathbf{R}, f(x, y) = 1 - e^{-}$ $(x^2 + y^2)$ *and* $\Gamma^h_{ij} = 0, i, j, h = 1, 2$. *Then* $\frac{\partial f}{\partial x} = 2xe^{-(x^2+y^2)}$, $\frac{\partial f}{\partial y} = 2ye^{-(x^2+y^2)}$ *and* f *has a unique critical* *minimum point* $(0,0)$. *However, the Euclidean Hessian of* f *is not positive semi-definite overall. We make previous reason for* $\sigma_{ij} = 2e^{-(x^2+y^2)}\delta_{ij}, a^1 = \frac{\partial f}{\partial x}, a^2 = \frac{\partial f}{\partial y}$ *Hence we obtain* $\overline{\Gamma}^h_{ij} = T^h_{ij}$,

$$\overline{\Gamma}^1_{11} = -\frac{2x^3}{x^2+y^2}, \overline{\Gamma}^1_{12} = \overline{\Gamma}^1_{21} = -\frac{2x^2y}{x^2+y^2}, \overline{\Gamma}^1_{22} = -\frac{2xy^2}{x^2+y^2},$$

$$\overline{\Gamma}^2_{11} = -\frac{2x^2y}{x^2+y^2}, \overline{\Gamma}^2_{12} = \overline{\Gamma}^2_{21} = -\frac{2xy^2}{x^2+y^2}, \overline{\Gamma}^2_{22} = -\frac{2y^3}{x^2+y^2}.$$

Observe that $\lim_{(x,y)\to(0,0)} T^h_{ij}(x,y) = 0$. *Hence take* $\overline{\Gamma}^h_{ij}(0,0) = 0$.

The next example shows what happens if we come out of the conditions of the previous theorem.

Example 1.3 *Let us take the function* $f : \mathbb{R} \to \mathbb{R}, f(x) = x^3$, *where the critical point* $x = 0$ *is an inflection point. We take* $\Gamma(x) = -1 - \frac{2}{x^2}$, *which is not defined at the critical point* $x = 0$, *but the relation of convexity is realized by prolongation,*

$$\sigma(x) = f''(x) - \Gamma(x)f'(x) = 3(x^2 + 2x + 2) > 0, \quad \forall x \in \mathbb{R}.$$

Let us consider the ODE of auto-parallels

$$x''(t) - \left(1 + \frac{2}{t^2}\right)x'(t)^2 = 0, t \neq 0.$$

The solutions

$$x(t) = -\frac{1}{2}\ln|-2 + t^2 - ct| + \frac{c}{\sqrt{8+c^2}} \operatorname{arctanh} \frac{2t - c}{\sqrt{8+c^2}} + c_1$$

are auto-parallels on $(\mathbb{R}\setminus\{0, t_1, t_2\}, \Gamma)$, *where* t_1, t_2 *are real solutions of* $-2 + t^2 - ct = 0$. *These curves are extended at* $t = 0$ *by continuity. The manifold* $(\mathbb{R}, \Gamma)$ *is not auto-parallely complete. Since the image* $x(\mathbb{R})$ *is not a "segment", the function* $f : \mathbb{R} \to \mathbb{R}, f(x) = x^3$ *is not globally convex.*

Remark 1.3 *For* $n \geq 2$, *there exists* C^1 *functions* $\varphi : \mathbb{R}^n \to \mathbb{R}$ *which have two minimum points without having another extremum point. As example,*

$$\varphi(x^1, x^2) = \left(x^{1^2} - 1\right)^2 + \left(x^{1^2}x^2 - x^1 - 1\right)^2$$

has two (global) minimum points $p = (-1, 0), q = (1, 2)$.

The restriction

$$\varphi(x^1, x^2) = \left(x^{1^4} + x^{1^4}x^{2^2} + 2x^1 + 2\right) - \left(x^{1^2} + 2x^{1^3}x^2 + 2x^{1^2}x^2\right), x^1 > 0, x^2 > 0$$

is difference of two affine convex functions (see Section 2).

Our chapter is based also on some ideas in: [3] (convex mappings between Riemannian manifolds), [7] (geometric modeling in probability and statistics), [13] (arc length in metric and Finsler manifolds), [14] (applications of Hahn-Banach principle to moment and optimization problems), [21] (geodesic connectedness of semi-Riemannian manifolds), and [28] (tangent and cotangent bundles). For algorithms, we recommend the paper [20] (sequential and parallel algorithms).

2. Optimizations with autoparallel restrictions

2.1. Direct theory

The auto-parallel curves $x(t)$ on the affine manifold (M, Γ) are solutions of the second order ODE system

$$\ddot{x}^h(t) + \Gamma^h_{ij}(x(t))\dot{x}^i(t)\dot{x}^j(t) = 0, x(t_0) = x_0, \dot{x}(t_0) = \xi_0.$$

Obviously, the complete notation is $x(t, x_0, \xi_0)$, with

$$x(t_0, x_0, \xi_0) = x_0, \dot{x}(t_0, x_0, \xi_0) = \xi_0.$$

Definition 2.1 *Let $D \subset M$ be open and connected and $f : D \to \mathbb{R}$ a C^2 function. The point $x_0 \in D$ is called minimum (maximum) point of f conditioned by the auto-parallel system, together with initial conditions, if for the maximal solution $x(t, x_0, \xi_0) : I \to D$, there exists a neighborhood I_{t_0} of t_0 such that*

$$f(x(t, x_0, \xi_0)) \geq \ (\leq)\ f(x_0), \quad \forall t \in I_{t_0} \subset I.$$

Theorem 2.1 *If $x_0 \in D$ is an extremum point of f conditioned by the previous second order system, then $df(x_0)(\xi_0) = 0$.*

Definition 2.2 *The points $x \in D$ which are solutions of the equation $df(x)(\xi) = 0$ are called critical points of f conditioned by the previous spray.*

Theorem 2.2 *If $x_0 \in D$ is a conditioned critical point of the function $f : D \to \mathbb{R}$ of class C^2 constrained by the previous auto-parallel system and if the number*

$$(\mathrm{Hess} f)_{ij}\, \xi_0^i \xi_0^j = \left(\frac{\partial^2 f}{\partial x^i \partial x^j} - \frac{\partial f}{\partial x^h}\, \Gamma^h_{ij} \right)(x_0)\, \xi_0^i \xi_0^j$$

is strictly positive (negative), then x_0 is a minimum (maximum) point of f constrained by the auto-parallel system.

Example 2.1 *We compute the Christoffel symbols on the unit sphere S^2, using spherical coordinates (θ, φ) and the Riemannian metric*

$$g_{\theta\theta} = 1, g_{\theta\varphi} = g_{\varphi\theta} = 0, g_{\varphi\varphi} = \sin^2\theta.$$

When $\theta \neq 0, \pi$, we find

$$\Gamma^{\theta}_{\varphi\varphi} = -\frac{1}{2}\sin 2\theta, \Gamma^{\varphi}_{\varphi\theta} = \Gamma^{\varphi}_{\theta\varphi} = \cot\theta,$$

and all the other Γs are equal to zero. We can show that the apparent singularity at $\theta = 0, \pi$ can be removed by a better choice of coordinates at the poles of the sphere. Thus, the above affine connection extends to the whole sphere.

The second order system defining auto-parallel curves (geodesics) on S^2 are

$$\ddot{\theta}(t) - \frac{1}{2}\sin 2\theta(t)\,\dot{\varphi}(t)\dot{\varphi}(t) = 0, \ddot{\varphi}(t) - 2\cot\theta(t)\,\dot{\varphi}(t)\dot{\theta}(t) = 0.$$

The solutions are great circles on the sphere. For example, $\theta = \alpha\, t + \beta$ and φ = const.

We compute the curvature tensor R of the unit sphere S^2. Since there are only two independent coordinates, all the non-zero components of curvature tensor R are given by $R^i_j = R^i_{j\theta\varphi} = -R^i_{j\varphi\theta}$, where $i, j = \theta, \varphi$. We get $R^{\theta}_{\varphi} = \sin^2\theta, R^{\varphi}_{\theta} = -1$ and the other components are 0.

Let $(\theta(t, (\theta_0, \varphi_0), \xi), \varphi(t, (\theta_0, \varphi_0), \xi), t \in R$ be the maximal auto-parallel which satisfies $\theta(t_0, (\theta_0, \varphi_0), \xi) = \theta_0$, $\dot{\theta}(t_0, (\theta_0, \varphi_0), \xi) = \xi^1$; $\varphi(t_0, (\theta_0, \varphi_0), \xi) = \varphi_0$, $\dot{\varphi}(t_0, (\theta_0, \varphi_0), \xi) = \xi^2$. We wish to compute $\min f(\theta, \varphi) = R^{\theta}_{\varphi} = \sin^2\theta$ with the restriction $(\theta(t, (\theta_0, \varphi_0), \xi), \varphi(t, (\theta_0, \varphi_0), \xi)), t \in R$.

Since $df = (2\sin\theta\cos\theta, 0)$, the critical point condition $df(\theta, \varphi)(\xi) = 0$ becomes $\sin\theta\cos\theta\,\xi^1 = 0$. Consequently, the critical points are either $(\theta_0 = k\pi, k \in \mathbb{Z}; \varphi)$, $(\xi^1, \xi^2) \neq (0, 0)$, or $(\theta_1 = (2k+1)\frac{\pi}{2}, k \in \mathbb{Z}; \varphi)$, $(\xi^1, \xi^2) \neq (0, 0)$, or $(\theta; \varphi)$, $(\xi^1 = 0; \xi^2 \neq 0)$.

The components of the Hessian of f are

$$(Hess f)_{\theta\theta} = \frac{\partial^2 f}{\partial\theta\partial\theta} = 2\cos 2\theta, (Hess f)_{\theta\varphi} = 0, (Hess f)_{\varphi\varphi} = \frac{1}{2}\sin^2 2\theta.$$

At the critical points (θ_0, φ) or (θ_1, φ), the Hessian of f is positive or negative semi-definite. On the other hand, along $(\xi^1 = 0, \xi^2 \neq 0)$, we find $(Hess f)_{ij}\,\xi^i\xi^j = \frac{1}{2}\sin^2 2\theta\,(\xi^2)^2 > 0, \xi^2 \neq 0$. Consequently, each point $(\theta \neq \frac{k\pi}{2}, \varphi)$, is a minimum point of f along each auto-parallel, starting from given point and tangent to $(\xi^1 = 0, \xi^2 \neq 0)$.

2.2. Theory via the associated spray

This point of view regarding extrema comes from paper [22].

The second order system of auto-parallels induces a spray (special vector field) $Y(x, y) = \left(y^h, \Gamma^h_{ij}(x)y^iy^j\right)$ on the tangent bundle TM, that is,

$$\dot{x}^h(t) = y^h(t), \dot{y}^h(t) + \Gamma^h_{ij}(x(t))y^i(t)y^j(t) = 0.$$

The solutions $\gamma(t) = (x(t), y(t)) : I \to D$ of class C^2 are called field lines of Y. They depend on the initial condition $\gamma(t)|_{t=t_0} = \left(x_0, y_0\right)$, and therefore the notation $\gamma\left(t, x_0, y_0\right)$ is more suggestive.

Definition 2.3 *Let $D \subset TM$ be open and connected and $f : D \to \mathbb{R}$ a C^2 function. The point $\left(x_0, y_0\right) \in D$ is called minimum (maximum) point of f conditioned by the previous spray, if for the maximal field line $\gamma\left(t, x_0, y_0\right)$, $t \in I$, there exists a neighborhood I_{t_0} of t_0 such that*

$$f\left(\gamma\left(t, x_0, y_0\right)\right) \geq \ (\leq) \ f\left(x_0, y_0\right), \quad \forall t \in I_{t_0} \subset I.$$

Theorem 2.3 *If $\left(x_0, y_0\right) \in D$ is an extremum point of f conditioned by the previous spray, then $\left(x_0, y_0\right)$ is a point where Y is in* Ker *df.*

Definition 2.4 *The points $(x, y) \in D$ which are solutions of the equation*

$$D_Y f(x, y) = df(Y)(x, y) = 0$$

are called critical points of f conditioned by the previous spray.

Theorem 2.4 *If $\left(x_0, y_0\right) \in D$ is a conditioned critical point of the function $f : D \to \mathbb{R}$ of class C^2 constrained by the previous spray and if the number*

$$\left(d^2 f(Y, Y) + df(D_Y Y)\right)\left(x_0, y_0\right)$$

is strictly positive (negative), then $\left(x_0, y_0\right)$ is a minimum (maximum) point of f constrained by the spray.

Example 2.2 *We consider the Volterra-Hamilton ODE system* [2].

$$\frac{dx^1}{dt}(t) = y^1(t), \frac{dx^2}{dt}(t) = y^2(t),$$

$$\frac{dy^1}{dt}(t) = \lambda y^1(t) - \alpha_1 y^{1^2}(t) - 2\alpha_2 y^1(t) y^2(t),$$

$$\frac{dy^2}{dt}(t) = \lambda y^1(t) - \beta_1 y^{2^2}(t) - 2\beta_2 y^1(t) y^2(t),$$

which models production in a Gause-Witt 2-species evolving in $\mathbb{R}^4$: *(1) competition if* $\alpha_1 > 0$, $\alpha_2 > 0$, $\beta_1 > 0$, $\beta_2 > 0$ *and (2) parasitism if* $\alpha_1 > 0$, $\alpha_2 < 0$, $\beta_1 > 0$, $\beta_2 > 0$.

Changing the real parameter t into an affine parameter s, we find the connection with constant coefficients

$$\Gamma^1_{11} = \frac{1}{3}(\alpha_1 - 2\beta_2), \Gamma^2_{22} = \frac{1}{3}(\beta_1 - 2\alpha_2),$$

$$\Gamma^1_{12} = \frac{1}{3}(2\alpha_2 - \beta_1), \Gamma^2_{12} = \frac{1}{3}(2\beta_2 - \alpha_1).$$

Let $x(t, x_0, y_0), t \in I$ *be the maximal field line which satisfies* $x(t_0, x_0, y_0) = (x_0, y_0)$. *We wish to compute* $\max f(x^1, x^2, y^1, y^2) = y^2$ *with the restriction* $x = x(t, x_0, y_0)$.

We apply the previous theory. Introduce the vector field

$$Y = \left(y^1, y^2, \lambda y^1 - \alpha_1 y^{1^2} - 2\alpha_2 y^1 y^2, \lambda y^1 - \beta_1 y^{2^2} - 2\beta_2 y^1 y^2\right).$$

We set the critical point condition $df(Y) = 0$. *Since* $df = (0, 0, 0, 1)$, *it follows the relation* $\lambda y^1 - \beta_1 y^{2^2}$ $-2\beta_2 y^1 y^2 = 0$, *that is, the critical point set is a conic in* $y^1 O y^2$.

Since $d^2 f = 0$, *the sufficiency condition is reduced to* $df(D_Y Y)(x_0, y_0) < 0$, *that is,*

$$\left(\lambda - \frac{\alpha_1 \beta_1 y^{22}}{\lambda - 2\beta_2 y^2} - 2\alpha_2 y^2\right)(y_0) < 0.$$

This last relation is equivalent either to

$$(\lambda - 2\alpha_2 y_0^2)(\lambda - 2\beta_2 y_0^2) - \alpha_1 \beta_1 y_0^2 2 < 0, \lambda - 2\beta_2 y_0^2 > 0$$

or to

$$(\lambda - 2\alpha_2 y_0^2)(\lambda - 2\beta_2 y_0^2) - \alpha_1 \beta_1 y_0^2 2 > 0, \lambda - 2\beta_2 y_0^2 < 0.$$

Each critical point satisfying one of the last two conditions is a maximum point.

3. Affine convexity of posynomial functions

For the general theory regarding geometric programming (based on posynomial, signomial functions, etc.), see [11].

http://dx.doi.org/10.5772/intechopen.75643

Theorem 3.1 *Each posynomial function is affine convex, with respect to some affine connection.*

Proof. A posynomial function has the form

$$f : \mathbb{R}^n_{++} \to \mathbb{R},\ f(x) = \sum_{k=1}^{K} c_k \prod_{i=1}^{n} \left(x^i\right)^{a_{ik}},$$

where all the coefficients c_k are positive real numbers, and the exponents a_{ik} are real numbers. Let us consider the auto-parallel curves of the form

$$\gamma(t) = \left(\left(a^1\right)^{1-t}\left(b^1\right)^t, \left(a^2\right)^{1-t}\left(b^2\right)^t, \dots, \left(a^n\right)^{1-t}\left(b^n\right)^t\right), t \in [0, 1],$$

joining the points $a = \left(a^1, \dots, a^n\right)$ and $b = \left(b^1, \dots, b^n\right)$, which fix, as example, the affine connection

$$\Gamma^h_{hj} = \Gamma^h_{jh} = -\frac{1}{2}\frac{\mu^h}{\mu^j x^j}, \text{ and otherwise } \Gamma^h_{ij} = 0.$$

It follows

$$f(\gamma(t)) = \sum_{k=1}^{K} c_k \prod_{i=1}^{n} \left(\left(a^i\right)^{a_{ik}}\right)^{1-t} \left(\left(b^i\right)^{a_{ik}}\right)^t$$

$$= \sum_{k=1}^{K} c_k \left(\prod_{i=1}^{n} \left(a^i\right)^{a_{ik}}\right)^{1-t} \left(\prod_{i=1}^{n} \left(b^i\right)^{a_{ik}}\right)^t.$$

One term in this sum is of the form $\psi_k(t) = A_k^{1-t} B_k^t$, and hence $\ddot{\psi}_k(t) = A_k^{1-t} B_k^t (\ln A_k - \ln B_k)^2 > 0$.

Remark 3.1 *Posynomial functions belong to the class of functions satisfying the statement "product of two convex function is convex".*

Corollary 3.1 *Each signomial function is difference of two affine convex posynomials, with respect to some affine connection.*

Proof. A signomial function has the form

$$f : \mathbb{R}^n_{++} \to \mathbb{R},\ f(x) = \sum_{k=1}^{K} c_k \prod_{i=1}^{n} \left(x^i\right)^{a_{ik}},$$

where all the exponents a_{ik} are real numbers and the coefficients c_k are either positive or negative. Without loss of generality, suppose that for $k = 1, \dots, k_0$ we have $c_k > 0$ and for $k = k_0 + 1, \dots, K$ we have $c_k < 0$. We use the decomposition

$$f(x) = \sum_{k=1}^{k_0} c_k \prod_{i=1}^{n} \left(x^i\right)^{a_{ik}} - \sum_{k=k_0+1}^{K} |c_k| \prod_{i=1}^{n} \left(x^i\right)^{a_{ik}},$$

we apply the Theorem and the implication $u''(t) \geq v''(t) \Rightarrow u - v$ convex. □

Corollary 3.2 *(1) The polynomial functions with positive coefficients, restricted to* $\mathbb{R}^n_{++}$*, are affine convex functions.*

(2) The polynomial functions with positive and negative terms, restricted to $\mathbb{R}^n_{++}$*, are differences of two affine convex functions.*

Proudnikov [18] gives the necessary and sufficient conditions for representing Lipschitz multivariable function as a difference of two convex functions. An algorithm and a geometric interpretation of this representation are also given. The outcome of this algorithm is a sequence of pairs of convex functions that converge uniformly to a pair of convex functions if the conditions of the formulated theorems are satisfied.

4. Bilevel disjunctive problem

Let $\left(M_1, {}^1\Gamma\right)$, the *leader decision affine manifold*, and $\left(M_2, {}^2\Gamma\right)$, the *follower decision affine manifold*, be two connected affine manifolds of dimension n_1 and n_2, respectively. Moreover, $\left(M_2, {}^2\Gamma\right)$ is supposed to be complete. Let also $f : M_1 \times M_2 \to \mathbb{R}$ be the *leader objective function*, and let $F = (F_1, \ldots, F_r) : M_1 \times M_2 \to \mathbb{R}^r$ be the *follower multiobjective function*.

The components $F_i : M_1 \times M_2 \to \mathbb{R}$ are (possibly) conflicting objective functions.

A *bilevel optimization problem* means a decision of leader with regard to a multi-objective optimum of the follower (in fact, a constrained optimization problem whose constraints are obtained from optimization problems). For details, see [5, 10, 12].

Let $x \in M_1$, $y \in M_2$ be the generic points. In this chapter, the *disjunctive solution set of a follower multiobjective optimization problem* is defined by

(1) the set-valued function

$$\psi : M_1 \rightrightarrows M_2, \psi(x) = \text{Argmin}_{y \in M_2} F(x, y),$$

where

$$\text{Argmin}_{y \in M_2} F(x, y) := \cup_{i=1}^{r} \text{Argmin}_{y \in M_2} F_i(x, y)$$

or

(2) the set-valued function

$$\psi : M_1 \rightrightarrows M_2, \psi(x) = \text{Argmax}_{y \in M_2} F(x, y),$$

where

$$\text{Argmax}_{y \in M_2} F(x, y) := \cup_{i=1}^{r} \text{Argmax}_{y \in M_2} F_i(x, y).$$

We deal with two bilevel problems:

(1) The *optimistic bilevel disjunctive problem*

$$(OBDP) \ \min_{x \in M_1} \ \min_{y \in \psi(x)} f(x, y).$$

In this case, the follower cooperates with the leader; that is, for each $x \in M_1$, the follower chooses among all its disjunctive solutions (his best responses) one which is the best for the leader (assuming that such a solution exists).

(2) The *pessimistic bilevel disjunctive problem*

$$(PBDP) \ \min_{x \in M_1} \ \max_{y \in \psi(x)} f(x, y).$$

In this case, there is no cooperation between the leader and the follower, and the leader expects the worst scenario; that is, for each $x \in M_1$, the follower may choose among all its disjunctive solutions (his best responses) one which is unfavorable for the leader.

So, a general optimization problem becomes a pessimistic bilevel problem.

Theorem 4.1 *The value*

$$\min_x [f(x, y) : y \in \psi(x)]$$

exists if and only if, for an index i, the minimum $\min_x [f(x, y) : y \in \psi_i(x)]$ *exists and, for each* $j \neq i$, *either* $\min_x \left[f(x, y) : y \in \psi_j(x)\right]$ *exists or* $\psi_j = \emptyset$. *In this case,*

$$\min_x [f(x, y) : y \in \psi(x)]$$

coincides to the minimum of minima that exist.

Proof. Let us consider the multi-functions $\phi_i(x) = f(x, \psi_i(x))$ and $\phi(x) = f(x, \psi(x))$. Then $\phi(x) = \cup_{i=1}^{k} \phi_i(x)$. It follows that $\min_x \phi(x)$ exists if and only if either $\min_x \phi_i(x)$ exists or $\psi_i = \emptyset$, and at least one minimum exists.

Taking minimum of minima that exist, we find

$$\min_x [f(x, y) : y \in \psi(x)]. \qquad \square$$

Theorem 4.2 *Suppose* M_1 *is a compact manifold. If for each* $x \in M_1$, *at least one partial function* $y \to F_i(x, y)$ *is affine convex and has a critical point, then the problem (OBDP) has a solution.*

Proof. In our hypothesis, the set $\psi(x)$ is nonvoid, for any x, and the compacity assures the existence of $\min_x f(x, \psi(x))$.

In the next Theorem, we shall use the Value Function Method or Utility Function Method. □

Theorem 4.3 *If a* C^1 *increasing scalarization partial function*

$$y \to L(x, y) = u(F_1(x, y), \ldots, F_k(x, y))$$

has a minimum, then there exists an index i such that $\psi_i(x) \neq \emptyset$. *Moreover, if* $f(x, y)$ *is bounded, then the bilevel problem*

$$\min_x [f(x, y) : y \in \psi(x)]$$

has solution.

Proof. Let $\min_y L(x, y) = L(x, y^*)$. Suppose that for each $i = 1, \ldots, k$, $\min_y F_i(x, y) < F_i(x, y^*)$. Then y^* would not be minimum point for the partial function $y \to L(x, y)$. Hence, there exists an index i such that $y^* \in \psi_i(x)$. □

Boundedness of f implies that the bilevel problem has solution once it is well-posed, but the fact that the problem is well-posed is shown in the first part of the proof.

4.1. Bilevel disjunctive programming algorithm

An important concept for making wise tradeoffs among competing objectives is bilevel disjunctive programming optimality, on affine manifolds, introduced in this chapter.

We present an exact algorithm for obtaining the bilevel disjunctive solutions to the multi-objective optimization in the following section.

Step 1: Solve

$$\psi_i(x) = \text{Argmin}_{y \in M_2} F_i(x, y), i = 1, \ldots, m.$$

Let $\psi(x) = \cup_{i=1}^{r} \psi_i(x)$ be a subset in M_2 representing the mapping of optimal solutions for the follower multi-objective function.

Step 2: Build the mapping $f(x, \psi(x))$.

Step 3: Solve the leader's following program

$$\min_x [f(x, y), y \in \psi(x)].$$

From numerical point of view, we can use the Newton algorithm for optimization on affine manifolds, which is given in [19].

5. Models of bilevel disjunctive programming problems

The manifold M is understood from the context. The connection Γ^h_{ij} can be realized in each case, imposing convexity conditions.

Example 5.1 *Let us solve the problem (cite* [7], *p. 7;* [9]*):*

$$\min_{(x_1,x_2)} F(x_1,x_2,y) = (x_1 - y, x_2)$$

subject to

$$(x_1,x_2) \in Argmin_{(x_1,x_2)}\{(x_1,x_2)\,|\,y^2 - x_1^2 - x_2^2 \geq 0\},$$

$$1 + x_1 + x_2 \geq 0, \; -1 \leq x_1, x_2 \leq 1, 0 \leq y \leq 1.$$

Both the lower and the upper level optimization tasks have two objectives each. For a fixed y value, the feasible region of the lower-level problem is the area inside a circle with center at origin ($x_1 = x_2 = 0$) *and radius equal to y. The Pareto-optimal set for the lower-level optimization task, preserving a fixed y, is the bottom-left quarter of the circle,*

$$\{(x_1,x_2) \in \mathbb{R}^2\,|\,x_1^2 + x_2^2 = y^2, x_1 \leq 0, x_2 \leq 0\}.$$

The linear constraint in the upper level optimization task does not allow the entire quarter circle to be feasible for some y. Thus, at most a couple of points from the quarter circle belongs to the Pareto-optimal set of the overall problem. Eichfelder [8] reported the following Pareto-optimal set of solutions

$$A = \left\{(x_1,x_2,y) \in \mathbb{R}^3\,|\,x_1 = -1 - x_2, x_2 = -\frac{1}{2} \pm \frac{1}{2}\sqrt{2y^2 - 1}, y \in \left[\frac{1}{\sqrt{2}}, 1\right]\right\}.$$

The Pareto-optimal front in $F_1 - F_2$ *space can be written in parametric form*

$$\left\{(F_1,F_2) \in \mathbb{R}^2\,|\,F_1 = -1 - F_2 - t, F_2 = -\frac{1}{2} \pm \frac{1}{2}\sqrt{2t^2 - 1}, t \in \left[\frac{1}{\sqrt{2}}, 1\right]\right\}.$$

Example 5.2 *Consider the bilevel programming problem*

$$\min_x \left[(x - y)^2 + x^2 : -20 \leq x \leq 20, y \in \psi(x)\right],$$

where the set-valued function is

$$\psi(x) = Argmin_y[xy : -x - 1 \leqslant y \leqslant -x + 1].$$

Explicitly,

$$\psi(x) = \begin{cases} [-1,1] & if\ x = 0 \\ -x - 1 & if\ x > 0 \\ -x + 1 & if\ x < 0. \end{cases}$$

Since $F(x,y) = (x-y)^2 + x^2$*, we get*

$$F(x,\psi(x)) = \begin{cases} [0,1] & \text{if } x = 0 \\ (-2x-1)^2 + x^2 & \text{if } x > 0 \\ (-2x+1)^2 + x^2 & \text{if } x < 0. \end{cases}$$

on the regions where the functions are defined.

Taking into account $(-2x-1)^2 + x^2 > 0$ *and* $(-2x+1)^2 + x^2 > 0$*, it follows that* $(x^\circ, y^\circ) = (0,0)$ *is the unique optimistic optimal solution of the problem. Now, if the leader is not exactly enough in choosing his solution, then the real outcome of the problem has an objective function value above* 1 *which is far away from the optimistic optimal value zero.*

Example 5.3 *Let* $F(x,y) = (F_1(x,y), F_2(x,y))$ *and a Pareto disjunctive problem*

$$\psi(x) = Argmin_y\, F(x,y) = Argmin_y\, F_1(x,y) \cup Argmin_y\, F_2(x,y).$$

Then it appears a bilevel disjunctive programming problem of the form

$$\min_x\ [f(x,y), y \in \psi(x)].$$

This problem is interesting excepting the case $\psi(x) = \emptyset, \forall x$*. If* $y \to F_1(x,y)$ *and* $y \to F_2(x,y)$ *are convex functions, then* $\psi(x) \neq \emptyset$*.*

To write an example, we use

$$F_1(x,y) = [xy : -x-1 \leqslant y \leqslant -x+1],\ F_2(x,y) = \left[x^2 + y^2 : y \geqslant -x+1\right]$$

and we consider a bilevel disjunctive programming problem of the form

$$\min_x\ \left[(x-y)^2 + x^2 : -20 \le x \le 20, y \in \psi(x)\right],$$

with

$$\psi(x) = \psi_1(x) \cup \psi_2(x),$$

where

$$\psi_1(x) = Argmin_y [xy : -x-1 \leqslant y \leqslant -x+1] = \begin{cases} [-1,1] & \text{if } x = 0 \\ -x-1 & \text{if } x > 0 \\ -x+1 & \text{if } x < 0, \end{cases}$$

$$\psi_2(x) = Argmin_y \left[x^2 + y^2 : y \geqslant -x+1\right] = \begin{cases} -x+1 & \text{if } x \le 1 \\ 0 & \text{if } x > 1, \end{cases}$$

$$\psi(x) = \begin{cases} [-1,1] & \text{if } x = 0 \\ \{-x-1, -x+1\} & \text{if } 0 < x \le 1 \\ \{-x-1, 0\} & \text{if } x > 1 \\ -x+1 & \text{if } x < 0. \end{cases}$$

The objective $f(x,y) = (x-y)^2 + x^2$ *and the multi-function* $\psi(x)$ *produce a multi-function*

$$f(x,\psi(x)) = \begin{cases} [0,1] & \text{if} \quad x = 0 \\ \left\{(2x+1)^2 + x^2, (2x-1)^2 + x^2\right\} & \text{if} \quad 0 < x \leq 1 \\ \left\{(2x+1)^2 + x^2, 2x^2\right\} & \text{if} \quad x > 1 \\ (2x-1)^2 + x^2 & \text{if} \quad x < 0. \end{cases}$$

In context, we find the inferior envelope

$$y(x) = \begin{cases} 0 & \text{if} \quad x = 0 \\ -x+1 & \text{if} \quad 0 < x \leq 1 \\ 0 & \text{if} \quad x > 1 \\ -x+1 & \text{if} \quad x < 0 \end{cases}$$

and then

$$f(x,y(x)) = \begin{cases} 0 & \text{if} \quad x = 0 \\ (2x-1)^2 + x^2 & \text{if} \quad x \in (-\infty,0) \cup (0,1] \\ 2x^2 & \text{if} \quad x > 1. \end{cases}$$

Since $(2x-1)^2 + x^2 > 0$, *the unique optimal solution is* $(x^\circ, y^\circ) = (0,0)$.

If we consider only $\psi_1(x)$ *as active, then the unique optimal solution* $(0,0)$ *is maintained. If* $\psi_2(x)$ *is active, then the optimal solution is* $(0,1)$.

6. Properties of minimum functions

Let $(M_1, {}^1\Gamma)$, the *leader decision affine manifold*, and $(M_2, {}^2\Gamma)$, the *follower decision affine manifold*, be two connected affine manifolds of dimension n_1 and n_2, respectively. Starting from a function with two vector variables

$$\varphi : M_1 \times M_2 \to \mathbb{R}, (x,y) \to \varphi(x,y),$$

and taking the infimum after one variable, let say y, we build a function

$$f(x) = \inf_y \{\varphi(x,y) : y \in a(x)\},$$

which is called *minimum function*.

A *minimum function* is usually specified by a pointwise mapping a of the manifold M_1 in the subsets of a manifold M_2 and by a functional $\varphi(x,y)$ on $M_1 \times M_2$. In this context, some differential properties of such functions were previously examined in [4]. Now we add new properties related to increase and convexity ideas.

First we give a new proof to Brian White Theorem (see Mean Curvature Flow, p. 7, Internet 2017).

Theorem 6.1 *Suppose that M_1 is compact, $M_2 = [0, T]$ and $f : M_1 \times [0, T] \to \mathbb{R}$. Let $\phi(t) = \min_x f(x, t)$. If, for each x with $\phi(t) = f(x, t)$, we have $\frac{\partial f}{\partial t}(x, t) \geq 0$, then ϕ is an increasing function.*

Proof. We shall prove the statement in three steps.

(1) *If f is continuous, then ϕ is (uniformly) continuous.*

Indeed, f is continuous on the compact $M_1 \times [0, 1]$, hence uniformly continuous. So, for $\varepsilon > 0$ it exists $\delta > 0$ such that if $|t_1 - t_2| < \delta$, then $|f(x, t_1) - f(x, t_2)| < \varepsilon$, for any $x \in M_1$, or

$$-\varepsilon < f(x, t_1) - f(x, t_2) < \varepsilon$$

On one hand, if we put $\phi(t_1) = f(x_1, t_1)$ and $\phi(t_2) = f(x_2, t_2)$, then we have

$$f(x, t_1) > f(x, t_2) - \varepsilon \geq f(x_2, t_2) - \varepsilon.$$

Hence $\min_x f(x, t_1) \geq f(x_2, t_2) - \varepsilon$, so is $\phi(t_1) - \phi(t_2) \geq -\varepsilon$.

On the other hand,

$$f(x, t_2) + \varepsilon > f(x, t_1) \geq f(x_1, t_1).$$

Hence $\min_x f(x, t_2) + \varepsilon \geq f(x_1, t_1)$, so is $\phi(t_1) - \phi(t_2) \leq \varepsilon$.

Finally, $|\phi(t_1) - \phi(t_2)| \leq \varepsilon$, for $|t_1 - t_2| < \delta$, that is, ϕ is (uniformly) continuous.

(2) *Let us fix $t_0 \in (0; T]$. If $\phi(t_0) = f(x_0, t_0)$ and $\frac{\partial f}{\partial t}(x_0, t_0) \geq 0$, then it exists $\delta > 0$ such that $\phi(t) \leq \phi(t_0)$, for any $t \in (t_0 - \delta; t_0)$.*

Suppose $\frac{\partial f}{\partial t}(x_0, t_0) > 0$, it exists $\delta > 0$ such that $f(x_0, t) \leq f(x_0, t_0)$, for each $t \in (t_0 - \delta; t_0)$. It follows $\min_x f(x, t) \leq f(x_0, t) \leq f(x_0, t_0)$, and so is $\phi(t) \leq \phi(t_0)$.

If $\frac{\partial f}{\partial t}(x_0, t_0) = 0$, then we use $\overline{f}(x, t) = f(x, t) + \varepsilon t$, $\varepsilon > 0$. For $\overline{f}$, the above proof holds, and we take $\varepsilon \to 0$.

(3) *ϕ is an increasing function.*

Let $0 \leq a < b \leq T$ and note $A = \{t \in [a, b] \,|\, \phi(t) \leq \phi(b)\}$. A is not empty. If $\alpha = \inf A$, then, by the step (2), $\alpha < b$ and, by the step (1), $\alpha \in A$. If $\alpha > a$, we can use the step (2) for $t_0 = \alpha$ and it would result that α was not the lower bound of A. Hence $\alpha = a$ and $\phi(a) \leq \phi(b)$.

Remark The third step shows that a function having the properties (1) and (2) is increasing. For this the continuity is essential. Only property (2) is not enough. For example, the function defined by $\phi(t) = t$ on $[0, 1]$ and $\phi(t) = 1 - t$ on $(1, 2]$ has only the property (2), but it is not increasing on $[0, 2]$.

Remark Suppose that f is a C^2 function and $\min_x f(x, t) = f(x_0(t), t)$, where $x_0(t)$ is an interior point of M. Since $x_0(t)$ is a critical point, we have

$$\phi'(t) = \frac{\partial f}{\partial t}(x_0(t), t) + < \frac{\partial f}{\partial x}(x_0(t), t), x_0'(t) > = \frac{\partial f}{\partial t}(x_0(t), t) \geq 0.$$

Consequently, $\phi(t)$ is an increasing function. If M_1 has a nonvoid boundary, then the monotony extends by continuity (see also the evolution of an extremum problem).

□

Example 6.1 *The single-time perspective of a function* $f : \mathbb{R}^n \to \mathbb{R}$ *is the function* $g : \mathbb{R}^n \times \mathbb{R}_+ \to \mathbb{R}$, $g(x, t) = tf(x/t)$, $\operatorname{dom} g = \{(x, t) | x/t \in \operatorname{dom} f, t > 0\}$. *The single-time perspective g is convex if f is convex.*

The single-time perspective is an example verifying Theorem 7.1. Indeed, the critical point condition for g, in x, $\frac{\partial g}{\partial x} = 0$, *gives* $x = tx_0$, *where* x_0 *is a critical point of f. Consequently,* $\phi(t) = \min_x g(x, t) = tf(x_0)$. *On the other hand, in the minimum point, we have* $\frac{\partial g}{\partial t}(x, t) = f(x_0)$. *Then* $\phi(t)$ *is increasing if* $f(x_0) \geq 0$, *as in Theorem 4.1.*

Theorem 6.2 *Suppose that* M_1 *is compact and* $f : M_1 \times M_2 \to \mathbb{R}$. *Let* $\phi(y) = \min_x f(x, y)$. *If, for each x with* $\phi(y) = f(x, y)$, *we have* $\frac{\partial f}{\partial y^\alpha}(x, y) \geq 0$, *then* $\phi(y)$ *is a partially increasing function.*

Proof. Suppose that f is a C^2 function and $\min_x f(x, y) = f(x_0(y), y)$, where $x_0(y)$ is an interior point of M_1. Since $x_0(y)$ is a critical point, we have

$$\frac{\partial \phi}{\partial y^\alpha} = \frac{\partial f}{\partial y^\alpha}(x_0(y), y) + < \frac{\partial f}{\partial x}(x_0(y), y), \frac{\partial x_0}{\partial y^\alpha} > = \frac{\partial f}{\partial y^\alpha}(x_0(y), y) \geq 0.$$

Consequently, $\phi(y)$ is a partially increasing function. If M has a non-void boundary, then the monotony extends by continuity. □

Theorem 6.3 *Suppose that* M_1 *is compact and* $f : M_1 \times M_2 \to \mathbb{R}$. *Let* $\phi(y) = \min_x f(x, y)$. *If, for each x with* $\phi(y) = f(x, y)$, *we have* $d_y^2 f(x, y) \leq 0$, *then* $\phi(y)$ *is an affine concave function.*

Proof. Without loss of generality, we work on Euclidean case. Suppose that f is a C^2 function and $\min_x f(x, y) = f(x(y), y)$, where $x(y)$ is an interior point of M_1. Since $x(y)$ is a critical point, we must have

$$\frac{\partial f}{\partial x^i}(x(y), y) = 0.$$

Taking the partial derivative with respect to y^α and the scalar product with $\frac{\partial x^i}{\partial y^\beta}$ it follows

$$\frac{\partial^2 f}{\partial x^i \partial x^j} \frac{\partial x^j}{\partial y^\alpha} \frac{\partial x^i}{\partial y^\beta} + \frac{\partial^2 f}{\partial y^\alpha \partial x^i} \frac{\partial x^i}{\partial y^\beta} = 0.$$

On the other hand

$$d_y\phi(y) = d_y f(x(y),y) = \left(\frac{\partial f}{\partial x^i}\frac{\partial x^i}{\partial y^\alpha} + \frac{\partial f}{\partial y^\alpha}\right)dy^\alpha = \frac{\partial f}{\partial y^\alpha}dy^\alpha$$

$$d_y^2\phi(y) = \left(\frac{\partial^2 f}{\partial y^\alpha \partial x^i}\frac{\partial x^i}{\partial y^\beta} + \frac{\partial^2 f}{\partial y^\alpha \partial y^\beta}\right)dy^\alpha dy^\beta$$

$$= \left(-\frac{\partial^2 f}{\partial x^j \partial x^i}\frac{\partial x^i}{\partial y^\beta}\frac{\partial x^j}{\partial y^\alpha} + \frac{\partial^2 f}{\partial y^\alpha \partial y^\beta}\right)dy^\alpha dy^\beta \leq 0.$$

□

Theorem 6.4 *Let $f : M_1 \times M_2 \to \mathbb{R}$ be a C^2 function and*

$$\phi(y) = \min_x f(x,y) = f(x(y),y).$$

If the set $A = \{(x(y),y) : y \in M_2\}$ is affine convex and $f|_A$ is affine convex, then $\phi(y)$ is affine convex.

Proof. Suppose f is a C^2 function. At points $(x(y),y)$, we have

$$0 \leq d^2 f(x(y),y) = \left(\frac{\partial^2 f}{\partial x^i \partial x^j}\frac{\partial x^i}{\partial y^\alpha}\frac{\partial x^j}{\partial y^\beta} + 2\frac{\partial^2 f}{\partial x^i \partial y^\alpha}\frac{\partial x^i}{\partial y^\beta} + \frac{\partial^2 f}{\partial y^\alpha \partial y^\beta}\right)dy^\alpha dy^\beta$$

$$= \left(\frac{\partial^2 f}{\partial x^i \partial y^\alpha}\frac{\partial x^i}{\partial y^\beta} + \frac{\partial^2 f}{\partial y^\alpha \partial y^\beta}\right)dy^\alpha dy^\beta = d^2\phi(y).$$

Author details

Constantin Udriste[1]*, Henri Bonnel[2], Ionel Tevy[1] and Ali Sapeeh Rasheed[1]

*Address all correspondence to: udriste@mathem.pub.ro

1 Department of Mathematics and Informatics, Faculty of Applied Sciences, University Politehnica of Bucharest, Bucharest, Romania

2 ERIM, Université de la Nouvelle-Calédonie, Nouma Cédex, New Caledonia, France

References

[1] Aké LA, Sánchez M. Compact affine manifolds with precompact holonomy are geodesically complete. Journal of Mathematical Analysis and Applications. 2016;**436**(2):1369-1371. DOI: 10.1016/j.jmaa.2015.12.037

[2] Antonelli PL, Ingarden RS, Matsumoto M. The Theory of Sparays and Finsler Spaces with Applications in Physics and Biology. Netherlands: Kluwer Academic Publishers, Springer; 1993. pp. 97-132. DOI: 10.1007/978-94-015-8194-3-4

[3] Arsinte V, Bejenaru A. Convex mappings between Riemannian manifolds. Balkan Journal of Geometry and Its Applications. 2016;**21**(1):1-14

[4] Beresnev VV, Pshenichnyi BN. The differential properties of minimum functions. USSR Computational Mathematics and Mathematical Physics. 1974;**14**(3):101-113. DOI: 10.1016/0041-5553(74)90105-0

[5] Bonnel H, Todjihounde L, Udrişte C. Semivectorial bilevel optimization on Riemannian manifolds. Journal of Optimization Theory and Applications. 2015;**167**(2):464-486. DOI: 10.1007/s10957-015-0789-6

[6] Boyd S, Vandenberghe L. Convex Optimization. United Kingdom: Cambridge University Press; 2009. 716p

[7] Calin O, Udrişte C. Geometric Modeling in Probability and Statistics. New York: Springer; 2014. DOI: 10.1007/978-3-319-07779-6

[8] Combettes PL. Perspective functions: Properties, constructions, and examples. Set-Valued and Variational Analysis. Dordrecht: Springer Science+Business Media; 2017:1-18. DOI: 10.1007/s11228-017-0407-x

[9] Das P, Roy TK. Multi-objective geometric programming and its application in gravel box problem. Journal of Global Research in Computer Science. 2014;**5**(7):6-11

[10] Deb K, Sinha A. Solving bilevel multi-objective optimization problems using evolutionary algorithms: KanGAL Report Number 2008005; 2017

[11] Duffin RJ, Peterson EL, Zener C. Geometric Programming. New Jersey: John Wiley and Sons; 1967

[12] Eichfelder G. Solving nonlinear multiobjective bilevel optimization problems with coupled upper level constraints: Technical Report Preprint No. 320, Preprint-Series of the Institut für Angewandte Mathematik, University of Erlangen-Nuremberg, Germany; 2007

[13] Myers B. Arc length in metric and Finsler manifolds. Annals of Mathematics. 1938;**39**(2): 463-471. DOI: 10.2307/1968797

[14] Olteanu O, Udrişte C. Applications of Hahn-Banach principle to moment and optimization problems. Nonlinear Functional Analysis and its Applications. 2005;**10**(5):725-742

[15] Poston T, Stewart I. Catastrophe Theory and Its Applications. Pitman; 1977. 491p

[16] Pripoae CL. Affine differential invariants associated to convex functions. In: Proceedings of Conference of SSM, Cluj, 1998; Digital Data Publishing House, Cluj-Napoca; 1999. pp. 247-252

[17] Pripoae CL, Pripoae GT. Generalized convexity in the affine differential setting. In: The International Conference "Differential Geometry—Dynamical Systems" (DGDS); 10–13 October 2013; Bucharest-Romania: BSG Proceedings 21; 2013. pp. 156-166

[18] Proudnikov I. The necessary and sufficient conditions for representing Lipschitz multivariable function as a difference of two convex functions. arXiv preprint arXiv:1409.5081v4. 2014

[19] Rasheed AS, Udrişte C, Ţevy I. Barrier algorithm for optimization on affine manifolds. BSG Proceedings. 2017;**24**:47-63

[20] Strongin DRG, Sergeyev YD. Global Optimization with Non-Convex Constraints: Sequential and Parallel Algorithms. London: Springer; 2000. 704p. DOI: 10.1007/978-1-4615-4677-1

[21] Sánchez M. Geodesic connectedness of semi-Riemannian manifolds. arXiv preprint math/0005039. 2001;**47**:3085-3102. DOI: 10.1016/s0362-546x(01)00427-8

[22] Udrişte C, Dogaru O. Extrema conditioned by orbits. (in Romanian). Buletinul Institutului Politehnic Bucureşti, Seria Mecanică. 1989;**51**:3-9

[23] Udrişte C. Convex Functions and Optimization Methods on Riemannian Manifolds. Netherlands: Kluwer Academic Publishers; 1994 . 365p. DOI: 10.1007/978-94-015-8390-9

[24] Udrişte C. Sufficient decreasing on Riemannian manifolds. Balkan Journal of Geometry and Its Applications. 1996;**1**:111-123

[25] Udrişte C. Riemannian convexity in programming (II). Balkan Journal of Geometry and Its Applications. 1996;**1**(1):99-109

[26] Udrişte C. Optimization methods on Riemannian manifolds. In: Proceedings of (IRB) International Workshop; 8–12 August 1995; Monteroduni. Italy: Algebras, Groups and Geometries; 1997. pp. 339-359

[27] Udrişte C, Bercu G. Riemannian Hessian metrics. Analele Universitatii din Bucuresti. 2005;**55**:189-204

[28] Yano K, Ishihara S. Tangent and Cotangent Bundles: Differential Geometry. New York: Dekker; 1973. 423p